Understanding Digital Electronics

2nd Edition

*R. H. Warring and
Michael J. Sanfilippo*

TAB BOOKS Inc.
Blue Ridge Summit, PA

*This second edition is dedicated to my mother.
Her courage and inspiration have given me an example
of dedication and commitment that I shall always
remember, and I shall ever be grateful to her.*

—M. J. Sanfilippo

SECOND EDITION
FIRST PRINTING

Understanding Digital Electronics was originally published in 1984.
Printed by permission of Lutterworth Press.

Copyright © 1990, 1984 by **R.H. Warring**
Printed in the United States of America

Reproduction or publication of the content in any manner, without express permission of the publisher, is prohibited. The publisher takes no responsibility for the use of any of the materials or methods described in this book, or for the products thereof.

Library of Congress Cataloging-in-Publication Data

Warring, R. H. (Ronald Horace), 1920-1984
   Understanding digital electronics / by R.H. Warring and Michael J. Sanfilippo. — 2nd ed.
     p.   cm.
   ISBN 0-8306-9226-6   ISBN 0-8306-3226-3 (pbk.)
   1. Digital electronics.   I. Sanfilippo, M. J.   II. Title
TK7868.D5W37   1989
621.381—dc20   89-36603
                                                                CIP

TAB BOOKS Inc. offers software for sale. For information and a catalog, please contact TAB Software Department, Blue Ridge Summit, PA 17294-0850.

Questions regarding the content of this book should be addressed to:

   Reader Inquiry Branch
   TAB BOOKS Inc.
   Blue Ridge Summit, PA 17294-0214

Acquisitions Editor: Kimberly Tabor
Technical Editor: Alyson Grupp
Production: Katherine Brown

Contents

	Introduction	iv
1	**Basic Digital Concepts**	1

Analog Systems—Digital Systems—Digital Terminology—Binary Numbers—Truth Tables

2 Symbols and Switches — 14

Digital Logic Gates—Memory—Simple Switching Functions—Series and Parallel Working—Simple Electronic Switches—Improving Transistor Switch-Off Times—Diode Switching—Schottky Diodes—Unijunction Transistors—Thyristors—Bounce-Free Switches

3 Mathematical Logic (Boolean Algebra) — 30

Basic Logic—Solving Problems—Boolean Algebraic Theorems

4 Logic Circuit Devices — 47

Diode-Transistor Logic (DTL)—Resistor-Transistor Logic (RTL)—Direct-Coupled-Transistor Logic (DCTL)—Emitter-Coupled-Transistor Logic (ECTL)—Transistor-Transistor Logic (TTL)—MOSFETs—Complementary MOS (CMOS)—MOS Logic—Clocked MOS Circuits—Dynamic MOS Inverters—Dynamic MOS NAND Gates—Handling MOS Devices—Integrated Circuits and Minimization—Standard IC Gates—Multiple Gate ICs—IC Buffers—Schmitt Trigger—Complex ICs—Digital Families Compared

5 Flip-Flops and Memories — 71

RS Flip-Flops—D Flip-Flops—JK Flip-Flops—The JK Master-Slave Flip-Flop—Sample-and-Hold—Read-Only-Memory (ROM)—Random-Access Memory (RAM)—Registers

6 Number Systems 85
Binary Coded Decimals—Types of Codes—Parity Bits—Other Number Systems—Handling Fractions

7 Digital Clocks 97
Operational Amplifier Clocks—IC Oscillators—Monostable Multivibrators—Bistable Multivibrators—Crystal Controlled Oscillators—Sweep Generators—Schmitt Trigger—IC Digital Clock

8 Encoders and Decoders 108
Encoders—Decoders—Multiplexers—Demultiplexers—IC Decoders—1-of-16 Decoder/Demultiplexer—LED Readout—Display Drivers

9 Digital Adders 123
Binary Adders—Half-Adders—Full-Adders—Binary Subtractors—Serial Adder/Subtractor—Half and Full-Subtractors

10 Binary Counters 134
The Basic Ripple Counter—Reversible Counter—Decade Counter—Divide-by-N Counter—Synchronous Counters—Johnson Counter (Twisted Ring Counter)—IC Binary Counters—IC Synchronous Counters

11 Converters and Registers 146
Digital-to-Analog Converters (D/A)—Analog-to-Digital Converters (A/D)—Shift Registers—IC Shift Register—Dynamic MOS Shift Register

12 The Arithmetic Logic Unit (ALU) 157
Cascading ALUs—ALU Functions—Microprocessors

Appendices

A Binary/Decimal Equivalents 168
B Simplifying Digital Logic Circuitry 173
C Computer Programming 182

Index 184

Introduction

THE book you are about to read, the second edition of *Understanding Digital Electronics*, has been completely revised and is very much up to date. Digital electronics, however, is constantly changing. It requires continuous review, vigilant reading, and constant experimenting from all of us who are a part of this exciting, ever-widening area of study and want to remain knowledgeable about it.

When I first undertook this project, I looked at this as an opportunity to review another writer's work and to perhaps, in some small way, add newer bits (no pun intended) of information to it. As a published author myself (*Solid-State Electronics Theory with Experiments*, TAB Books, 1987) I knew the amount of effort required to research, write, and illustrate a technical book. In that sense, this book was very much a challenge.

The majority of the chapters in this second edition contain all new circuit diagrams and most contain information not found in the first edition. Existing diagrams were corrected to reflect electronic symbols used in this country. Most concepts were expanded upon and explained in simpler, layman's terms.

The most pronounced change is in the last chapter, "The Arithmetic Logic Unit (ALU)." The original book based the theory of the ALU on a device foreign to us here in the United States. I found it more appropriate to base ALU theory on the 74181, a device common to many electronic circuit labs in technical schools throughout the country.

Also, binary arithmetic, along with a great many other topics, was added to this second edition. This book is, therefore, considerably more complete and up-to-date than its predecessor. That is not to say that it is all-inclusive; no book ever is. It does reflect a somewhat simpler approach with more information.

My intentions, then, were to strive for a second edition which would be more enjoyable and easier to understand without sacrificing the good intent of the first edition author. I hope I have achieved those objectives and have instilled in you, the reader, a desire for further study of digital electronics and eventually, microprocessors.

1

Basic Digital Concepts

TO understand digital electronics, you first need to understand the terms *digital electronics* and *analog electronics*, as well as the basic differences between the two. This will help those of you with a knowledge of basic and/or solid-state electronics identify with digital concepts. If you don't have an electronics background, you will find that digital concepts are relatively easy to understand. This chapter deals with some of those concepts and shows the simplicity, and hence the beauty, of digital electronics.

We will explain some advantages and disadvantages of each type of system or concept, then introduce binary arithmetic, and finally provide some information about truth tables. Truth tables are an invaluable tool in designing and troubleshooting all kinds of digital electronic circuits, from the simplest to the most complex.

ANALOG SYSTEMS

Analog systems offer a faster response to changes in the analog input signal, and less distortion in systems designed to amplify and reproduce the input analog signal.

However, analog systems also involve higher power dissipation, more weight, larger size (usually translated to higher costs), and greater sensitivity to environmental, or temperature, changes.

DIGITAL SYSTEMS

Digital systems have many advantages, including smaller size, less power dissipation, lower costs, lighter weight, and less sensitivity to environmental changes.

Disadvantages of digital systems are few, and are becoming less and less as our manufacturing processes and engineering capabilities improve. The most significant are induced distortion, which is an inherent effect of converting an analog input signal into a digital output signal (*A to D* or *A/D conversion*); and a comparatively long response time in performing the opposite operation (*D to A* or *D/A conversion*). Today's technology has reduced that response time, but it is still a considered factor in electronic circuit design.

DIGITAL TERMINOLOGY

Digital electronics uses a relatively new vocabulary. It is a vocabulary filled with terms that are very logical. They also provide you with a useful base of electronic terms, which is helpful if you plan to study further those areas of electronics that include microprocessors and circuits that are controlled by microprocessors.

Some of these terms are *AND gates, OR gates, NAND gates, NOR gates,* and *NOT gates.* The most important of these are the AND, OR, and NOT gates. Almost all of today's digital computers operate on the concepts of these three simple digital electronic devices.

BINARY NUMBERS

Digital systems employ *binary devices* which function in only two states. (Binary simply means two.) These two states are described in various ways:

- on/off or close/open for switching devices
- 1/0 for counting or computing devices
- true/false or yes/no for logic devices
- pulse/no pulse for trigger circuits
- high/low for practical circuits where voltage levels are relative (i.e., a low signal is not necessarily zero)

All these pairs of terms mean the same thing, one state or the other, with no intermediate state. This is the whole basis of digital (binary) working.

Binary numbers are based on just two digits, 1 and 0. Individual digits in a binary number then represent equivalent powers of 2, instead of 10 as in the decimal system. A particular advantage of the binary system is that there are no multiplication tables as such, and any problem involving addition, subtraction, multiplication, or division can be broken down into a series of individual binary operations, with each switching element in the system being continuously used (that is, either in the on or off state).

Compared with the decimal system, binary numbers are tedious as a written language. For example, TABLE 1-1 shows the binary equivalents of decimal numbers from 1 to 32.

Remembering that the binary system is based on powers of 2, the simplest way to derive the binary equivalent of a large decimal number is to subtract the highest power of 2 contained by the number, then subtract the highest power of 2 from the remainder, and so on until only a 1 or 0 is left as the remainder. For example, to find the binary equivalent of the decimal number 269, perform the following operation:

The highest power of 2 within 269 is $2^8 = 256$.
This leaves $269 - 256 = 13$.
The highest power of 2 within this remainder is $2^3 = 8$.
This leaves $13 - 8 = 5$.

4 Basic Digital Concepts

Table 1-1. Decimal-To-Binary Conversion

Decimal	Binary Number	Decimal	Binary Number
1 (2^0)	1	17	10001
2 (2^1)	10	18	10010
3	11	19	10011
4 (2^2)	100	20	10100
5	101	21	10101
6	110	22	10110
7	111	23	10111
8 (2^3)	1000	24	11000
9	1001	25	11001
10	1010	26	11010
11	1011	27	11011
12	1100	28	11100
13	1101	29	11101
14	1110	30	11110
15	1111	31	11111
16 (2^4)	10000	32 (2^5) and so on	100000

The highest power of 2 within this remainder is $2^2 = 4$. This leaves $5 - 4 = 1$.

The corresponding number is thus $2^8 + 2^3 + 2^2$ with a remainder of 1. Another way to show it is:

Power	Binary		Decimal
$2^8 =$	100000000	=	256 decimal
$2^3 =$	1000	=	8 decimal
$2^2 =$	100	=	4 decimal
remainder =	1	=	1 decimal
	100001101	=	269 decimal

This binary number is long, consisting of 9 digits (bits). It counts in a system involving only 1 or 0 so it can readily be handled by digital devices. The number of *bits* (binary digits) to be handled in a calculation does not represent any practical limitation. The speed at which these devices can work is extremely high. Here, for example, is the number of bits different types of digital circuit devices can handle per second:

MOS(metal-oxide-semiconductor): 3-4 million
CMOS(complementary-metal-oxide-semiconductor): 10-15 million
HTL(high-threshold-logic): 20 million
DTL(diode-transistor-logic): 35 million
RTL(resistor-transistor-logic): 80 million
TTL(transistor-transistor-logic): 170 million
ECL(emitter-coupled-logic): 250-1000 million

Binary Arithmetic

Adding and subtracting can be performed in binary just as in decimal, but certain rules must be followed when using binary digits or bits:

Addition	Subtraction
$0+0 = 0$	$0-0 = 0$
$0+1 = 1$	$1-1 = 0$
$1+0 = 1$	$1-0 = 1$
$1+1 = 10$	
$1+1+1 = 11$	

As you can see, remembering these rules makes adding and subtracting binary numbers quite simple. An example of adding two binary numbers is shown below:

Binary	Decimal Equivalent
1011	11
+100	+4
1111	15

Here is an example of binary subtraction:

Binary	Decimal Equivalent
11011	27
−1001	−9
10010	18

In digital electronic circuits, devices called *half adders* and *full adders* perform binary addition and subtraction. Additional information on these types of circuits may be found in chapter 10.

Binary Coded Decimal

To simplify working with large numbers a hybrid system known as a *binary coded decimal* (BCD) is normally used. Here, separate groups of binary digits are used to express units, tens, hundreds, etc. Since each binary group needs to be able to accommodate a count of up to 9, it must consist of four digits; that is, to accommodate 9 it must run 1001. (Refer to TABLE 1-1.)

The number 269 (100001101 in the binary system) is, as a binary coded decimal:

0010 0110 1001

equivalent to

2 6 9

in decimal numbers.

For the next number up, 270, the right-hand binary group changes to 1010, representing 10, which is immediately carried forward into the next group. The binary coded decimal would then read:

0010 0111 0000

equivalent to

2 7 0

in decimal numbers.

Binary coded decimal systems are described further in chapter 7.

TRUTH TABLES

Truth tables are an easily-understood way to represent the way digital devices and circuits work. They are widely used with Boolean Algebra (discussed in chapter 4) to solve circuit design problems. A truth table lays out the complete range of signal states for a device in terms of 1 (signal on) or 0 (signal absent).

Starting with the simplest device, an inverter or NOT gate, there is one signal input, A (which may have a state of 0 or 1) and one signal output, X. An inverter makes the state of output X the inversion or opposite of input A. The truth table then reads as shown in TABLE 1-2. This fully expresses all of the possible working states (two in this case) of the inverter, sometimes referred to as a NOT logic element or gate.

Table 1-2. NOT Truth Table

A	X
0	1
1	0

All other devices have more than one input. A basic rule to follow here in compiling a truth table is that with series logic, all the inputs must be 1 before the output can be 1, and with parallel logic the output is always 1 if any of the inputs is 1. This is equally well explained by mechanical thinking, since parallel logic is the equivalent to a number of on-off switches connected in parallel (any one switch which is on will pass a signal) and series logic is equivalent to a number of on-off switches connected in series where all of the switches must be on before a signal can be passed. The basic truth tables for such devices, written for two inputs, are shown in TABLES 1-3 and 1-4.

A	B	X
0	0	0
1	0	0
0	1	0
1	1	1

Table 1-3. Series Logic

A	B	X
0	0	0
1	0	1
0	1	1
1	1	1

Table 1-4. Parallel Logic

An example of a truth table for an OR gate, sometimes called an OR logic element, with two inputs is shown in TABLE 1-5. It is, in fact, an example of a parallel logic device. Expanded to cover more than two inputs, the same basic rule applies. X equals 1 when any input equals 1. TABLE 1-6 illustrates the truth table for a four-input OR gate. There are, as you can see, sixteen different states possible with any four-input gate. In the case of the OR device, fifteen of these give the output signal.

A	B	X
0	0	0
1	0	1
0	1	1
1	1	1

Table 1-5. OR Truth Table

An example of series logic is the AND gate. Its basic truth table for a two-input device is shown in TABLE 1-7.

TABLE 1-8 is an illustration of the AND gate written out for four inputs. Again, there are sixteen possible different states, but only one provides an output of $X = 1$. With sixteen switches connected in series (in mechanical terms) the path through them from input to output remains broken until all the switches are on.

Table 1-6. Four-Input OR Gate

A	B	C	D	X
0	0	0	0	0
1	0	0	0	1
0	1	0	0	1
0	0	1	0	1
0	0	0	1	1
1	1	0	0	1
1	0	1	0	1
1	0	0	1	1
1	1	1	0	1
1	1	0	1	1
1	0	1	1	1
0	1	1	1	1
0	1	0	1	1
0	1	1	1	1
0	0	1	1	1
1	1	1	1	1

Table 1-7. AND Truth Table

A	B	X
0	0	0
1	0	0
0	1	0
1	1	1

Table 1-8. Four-Input AND Gate

A	B	C	D	X
0	0	0	0	0
1	0	0	0	0
0	1	0	0	0
0	0	1	0	0
0	0	0	1	0
1	1	0	0	0
1	0	1	0	0
1	0	0	1	0
1	1	1	0	0
1	1	0	1	0
1	0	1	1	0
0	1	0	1	0
0	1	1	1	0
0	0	1	1	0
1	1	1	1	1

10 Basic Digital Concepts

The immediate reaction to these two examples is probably a feeling that it is much simpler to work in terms of switching equivalents than truth tables—and for very simple problems in digital logic it is. However, most problems require a combination of logic devices to provide the solution, which may involve both series and parallel logic. Drawing out the switching circuits can then become a more elaborate process than plotting truth tables and be more susceptible to mistakes.

Truth tables for other logic gates are given below. These are written for two-input devices. They can be expanded to present truth tables for more than two inputs by following the same established pattern.

The truth table for a NOR logic gate with two inputs is shown in TABLE 1-9. This can be identified as *inverted series logic*. Note also that inversion has changed the parallel logic of the OR gate to series logic in the case of the NOR (Not OR) gate. The significance of this occurs frequently when working with Boolean Algebra.

A	B	X
0	0	1
1	0	0
0	1	0
1	1	0

Table 1-9. NOR Truth Table

The truth table for a NAND gate with two inputs is shown in TABLE 1-10. This can be identified as *inverted parallel logic*. Inversion has changed the series logic of AND to parallel logic in the case of NAND (Not AND).

A	B	X_1
0	0	0
1	0	1
0	1	1
1	1	1

Table 1-10. OR Truth Table

Combinations of Logic Gates

The state of combinations of logic gates can be expressed in the same way as a truth table. Suppose, for example, the design requirement is to provide for input signals A or B to produce an output signal only in combination with a third input signal C. (For example, A and B are trainee operators who can only give a command signal to a machine when the instructor (C) also adds his or her own signal.) FIGURE 1-1 shows how this can be implemented using OR and AND logic.

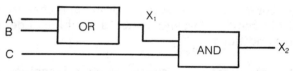

Fig. 1-1. A typical combination of logic gates.

TABLES 1-10 through 1-12 show how the truth tables are arranged beginning with the OR device or gate, proceeding to the AND device, and finally culminating in the truth table for the combination of both the OR and AND devices. Writing the truth table for the OR device first and calling the output X_1 is shown in TABLE 1-10. X_1 is now one of the inputs to the AND device. The truth table for this device is then shown in TABLE 1-11. The combined truth table can then be written as shown in TABLE 1-12. Note that the number of states is equal to the number of devices multiplied by the number of inputs to each device. In this example, $3 \times 2 = 6$ possible states.

Table 1-11. AND Truth Table

X_1	C	X_2
0	0	0
1	0	0
0	1	0
1	1	1

12 Basic Digital Concepts

A	B	C	X_2
0	0	0	0
1	0	0	0
0	1	1	1
1	1	0	0
1	1	1	1
1	0	1	1

Table 1-12. OR and AND Truth Table

Plotting Truth Tables

A truth table can be drawn up as a starting point in design. For example, suppose the problem is concerned with a control circuit to start and operate a machine under the following conditions:

A = signal from operator standing by machine
OR D = signal from a remote start
AND B = signal confirming guard is in place
AND C = signal from detector showing workpiece is in place

The machine must not start under any other conditions.

There are four inputs to consider, which results in sixteen possible combinations or states. This establishes the basis for writing out a five-column, sixteen-line truth table. On the output column, 1 must appear only when A = 1 OR D = 1 AND B = 1 AND C = 1. All the other combinations of A, B, C, and D must give X = 0. This is shown in TABLE 1-13.

FIGURE 1-2 shows this truth table implemented with logic devices and also with mechanical switches.

A	B	C	D	X
0	0	0	0	0
0	0	0	1	0
0	0	1	0	0
0	0	1	1	0
0	1	0	0	0
0	1	0	1	0
0	1	1	0	0
0	1	1	1	1
1	0	0	0	0
1	0	0	1	0
1	0	1	0	0
1	0	1	1	0
1	1	0	0	0
1	1	0	1	0
1	1	1	0	1
1	1	1	1	1

Table 1-13. Truth Table Used for Design

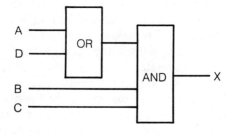

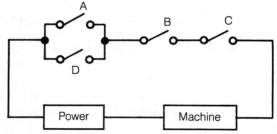

Fig. 1-2. *Logic solution (top) and implementation with mechanical switches (bottom).*

2

Symbols and Switches

ONE of the most confusing things about the use of symbols representing the various logic elements or gates is that the original (and literally logical) way of designating them in the form of annotated blocks has largely been abandoned in favor of representative symbols, the significance of which is not apparent until you are familiar with them. Even then, misunderstandings can easily arise, since over the years different symbols have been used to illustrate the same function(s). Various attempts have been made to standardize symbols; US MIL standard recommendations are used in American literature while CETOP standards are widely used in Europe.

Another source of confusion is that different letter symbols are used to designate inputs and outputs, particularly for basic devices. These include A, B, C...N for inputs and W, X, Y, Z for outputs. This is not particularly important if the application is clear, but can cause confusion with more complex devices where specific symbols (and sometimes different symbols) are used to designate specific integrated circuit (IC) terminals. Examples are Ck for clock input and D for data input.

DIGITAL LOGIC GATES

The simple block method of symbolizing logic elements is obvious, readily readable, and needs no extensive description. All symbols are in the form of a rectangular block with the function written inside. Input lines are added to the left side of the block and an output line to the right. FIGURE 2-1 shows a number of representative logic elements with two or more inputs and one output each. (The NOT gate has only one input.) For the sake of consistency, separate inputs are designated A, B, C, etc. The output line is designated X. The value 1 is used to designate a signal present, and a 0 represents no signal at that line.

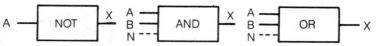

Fig. 2-1. *Examples of block logic symbols and annotation.*

Such block symbols are rarely used now, except in elementary textbooks, so you need to know the alternate forms of other basic symbols. Each function is dealt with separately and is illustrated with its corresponding mechanical switching function. (Devices are restricted to two inputs for simplicity.) The switching function is shown as 1 in the operated position and 0 in the off position. As you can see in the following drawings, both the standard gate symbol and, where appropriate, its mechanical switch equivalent are represented.

The YES Gate

The YES gate is a 1-input, 1-output device with input and output always the same; that is, $A=0$, $X=0$; or $A=1$, $X=1$, where A is the input signal and X is the output signal. The YES gate is also referred to as a *buffer*. The symbol for a YES gate is shown in FIG. 2-2.

Fig. 2-2. YES logic symbols.

The NOT Gate (Inverter)

The NOT gate, a 1-input, 1-output device, works the other way round to the YES gate. If there is an input to A, there is no output at X and vice versa (that is, A=1, X=0; or A=0, X=1). The symbols in FIG. 2-3 show this inverted mode of working by means of a circle on the output side.

Fig. 2-3. NOT logic symbols.

The AND Gate

The AND gate produces a logic level 1 at its output only when both inputs, A and B, are at logic level 1. (See FIG. 2-4.) In mechanical form, it is two switches in series.

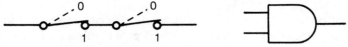

Fig. 2-4. AND logic symbols.

The NAND Gate

The NAND gate is an inverted form of the AND gate where there is a logic level 1 output for all input states except when both A=1 and B=1; in that case, X=0. (See FIG. 2-5.)

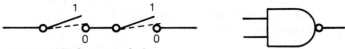

Fig. 2-5. NAND logic symbols.

The OR Gate

The OR gate is the equivalent of a parallel switching circuit. When either switch is closed, or both switches are closed, there is an output. (See FIG. 2-6.)

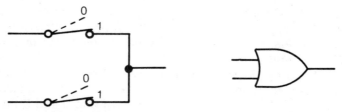

Fig. 2-6. OR logic symbols.

The NOR Gate

The NOR gate is the inverted form of the OR gate, so once again the symbols have the inversion mark (a circle on the output side) added, as shown in FIG. 2-7. Note that the switches are normally closed and there is an output ($X = 1$) only when both $A = 0$ and $B = 0$.

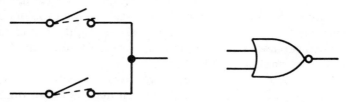

Fig. 2-7. NOR logic symbols.

The Exclusive OR (XOR) Gate

The exclusive OR gate is a special form of AND logic providing an output only when one particular input is equal to 1. If a 1 appears at the other input, it is inhibited or inverted to 0. Basically, in fact, this is an AND gate with one input inverted. FIGURE 2-8 shows the symbol for the exclusive OR (XOR) gate.

18 Symbols and Switches

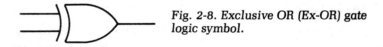

Fig. 2-8. Exclusive OR (Ex-OR) gate logic symbol.

MEMORY

Memory function is performed by a *flip-flop* (FF) which performs a store rather than a switching function. The output state depends on the last input applied and is maintained when the inputs are resumed.

In practice, there are different types of flip-flops, each of which is given its specific symbol and inputs and designated accordingly; that is R and S for an RS flip-flop, J and K for a JK flip-flop, D for a D type flip-flop, and T for a T type flip-flop. Outputs are then designated Q and $\overline{Q}$. In addition, the flip-flop may have a clock signal input (designated C, or Ck); a clear signal input (C); and a preset input (P) depending on type. These symbols are illustrated in FIG. 2-9. For more detailed information on flip-flops, see chapter 6.

SIMPLE SWITCHING FUNCTIONS

As an example of the application of logic to the design of switching circuits using digital devices, take the problem of designing a circuit for switching a single light on and off from two separate points. This is a common arrangement in the hallway or stairway of a house.

The basic requirements are two possible inputs (switches)—call them A and B—which may be either on or off. When either A or B is on, there is an output (that is, a circuit completed to light the bulb). A and B cannot be on at the same time. If one is on, operating the other switch switches the light off.

This can be written in the form of a truth table (TABLE 2-1). A 1 under columns A or B represents a switch on and a 1 under column L represents the light on. This can also be expressed in the form of this equation:

$$L = A\overline{B} + \overline{A}B$$

Simple Switching Functions 19

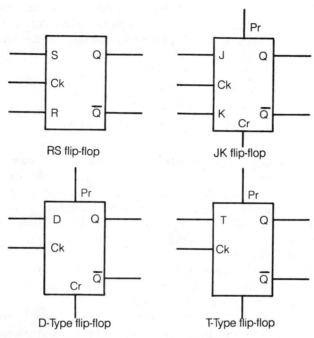

Fig. 2-9. Symbols for different types of flip-flops.

Table 2-1. 1 = sw. on, 1 = light on

A	B	L
0	0	0
0	1	1
1	0	1
1	1	0

This states L = A on AND B off OR A off AND B on. More about equations in the next chapter, but for now a further equation can be written expressing the combinations that do not produce an output; that is, do not switch the light on. This output can be represented by the letter D for dark and follows:

$$D = \overline{L} = AB + \overline{A}\overline{B}$$

The validity of the first equation can be proved using this second equation. Applying deMorgan's theorem (to be discussed in detail in the next chapter) this second statement becomes:

$$L = \overline{(AB + \overline{A}\overline{B})}$$
$$= (A + B)(\overline{A} + \overline{B})$$
$$= A\overline{A} + A\overline{B} + \overline{A}B + B\overline{B}$$
$$= A\overline{B} + \overline{A}B$$

This restates, and proves the validity of, the first formula. However, it also provides a second equation for implementing the requirements specifically in binary (on-off) elements to produce the desired switching circuit.

The first equation $L = \overline{A}B + A\overline{B}$ (FIG. 2-10) is implemented in terms of mechanical switches (or relay contacts) and also in terms of logic gates.

FIGURE 2-11 shows the second equation, $L = (A + B) \cdot (\overline{A} + \overline{B})$, implemented in terms of mechanical switches (or relays) and also in terms of logic gates.

FIGURE 2-10 is obviously the best practical solution, since it involves only half the contacts (series logic as opposed to

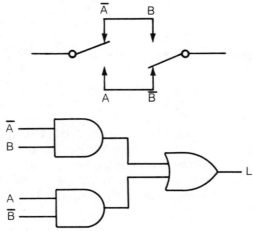

Fig. 2-10. First solution to switching problem.

Simple Switching Functions 21

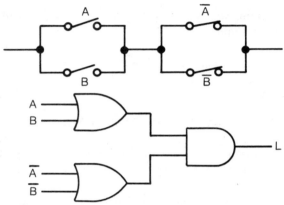

Fig. 2-11. Alternate solution to switching problem.

parallel logic). In the case of the gate solutions, the choice is not so obvious. It largely depends on the type gates most readily available. FIGURE 2-10 requires two AND gates and one OR gate. FIGURE 2-11 requires two OR gates and one AND gate.

These solutions may seem overcomplicated for the problem involved. Basically, they are presented to show the principle of digital switching circuit design with a simple, easily-understood example. Suppose we take it one step further to derive suitable circuitry for switching a light on from any of three different switch points.

The starting point is to draw up the truth table as shown in TABLE 2-2. This establishes all the possible input conditions, but does not give any immediate clue as to possible circuit design without drawing out each combination in detail.

Table 2-2. All Possible Combinations

A	B	C	L
0	0	0	0
0	0	1	1
0	1	0	1
0	1	1	0
1	0	0	1
1	0	1	0
1	1	0	0
1	1	1	1

You can derive a formula from the truth table (or original logic requirements):

$$L = A\overline{B}\overline{C} + \overline{A}B\overline{C} + \overline{A}\overline{B}C + ABC$$

This factors as follows:

$$L = \overline{C}(A\overline{B} + \overline{A}B) + C(\overline{A}\overline{B} + AB)$$

It is now possible to simplify to some extent by calling $A\overline{B} + \overline{A}B = X$. Then, since $A\overline{B} + \overline{A}B = \overline{AB + \overline{A}\overline{B}}$:

$$L = \overline{C}X + C\overline{X}$$

Solutions to this equation implemented in the form of both mechanical switches and exclusive OR logic gates are shown in FIG. 2-12.

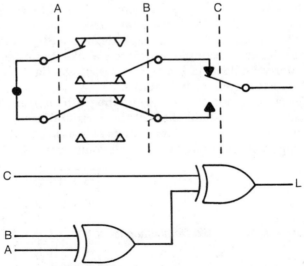

Fig. 2-12. Three-position light switching solution.

SERIES AND PARALLEL WORKING

The difference between digital devices operating in series or parallel modes is easy to explain diagramatically. In series operation, binary digits are expressed by voltage levels in a single output wire displaced in time. Thus a complete signal representing the binary number, say, 100101 is as shown in FIG. 2-13. (This is for positive working; it could equally well be given by negative working. In this case, the 0 level could

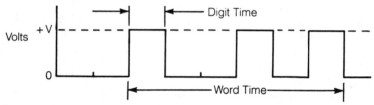

Fig. 2-13. Series working with positive logic (1 = +V).

be +V, with each pulse appearing as a 0 value; alternately, the 0 level could be a 0 value and each pulse level being −V.)

In parallel operation, each digit is allocated a separate line. Outputs then appear simultaneously on each line as shown in FIG. 2-14. Again, a negative instead of a positive voltage value could represent a 1. In many practical circuits, too, the change in voltage or signal swing may be from some nominal voltage representing condition 0 to some more positive (or more negative) voltage representing a 1. In such cases the description HIGH or H is commonly used to designate a 1 signal, and LOW or L represents a 0 signal. In other words, HIGH (or H) is used instead of 1; and LOW (or L) instead of 0.

Series working may appear the logical choice since it needs only one digital device or output wire to handle any number of digits. Parallel working has the disadvantage of requiring n devices or output wires to handle n digits, or n times as much circuit hardware to handle the same information. However, it has the advantage of being n times as fast as series working. In practical circuits, however, both the number of components used and operating time may be modified by other factors.

24 Symbols and Switches

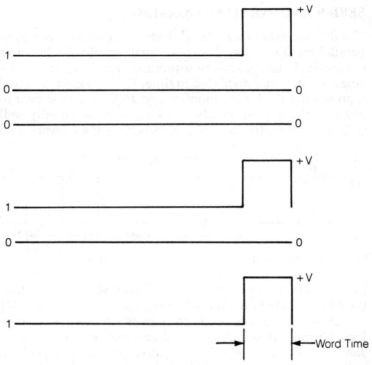

Fig. 2-14. Parallel working with positive logic (1 = +V).

SIMPLE ELECTRONIC SWITCHES

A bipolar junction transistor can readily work as a switch although its characteristics are not ideal for this purpose. The most usual way of working is in the saturation mode, when the transistor has two stable states, one passing no current (except for leakage current) corresponding to off, and the other in the saturated state passing maximum current and corresponding to on. (See FIG. 2-15.)

In the off condition, the collector voltage approaches V_{cc}. In the on condition, the collector voltage is V_{CE}, which is typically on the order of 0.15 to 0.6 volts. However, the transistor is now capable of passing a (relatively) large voltage.

Simple Electronic Switches 25

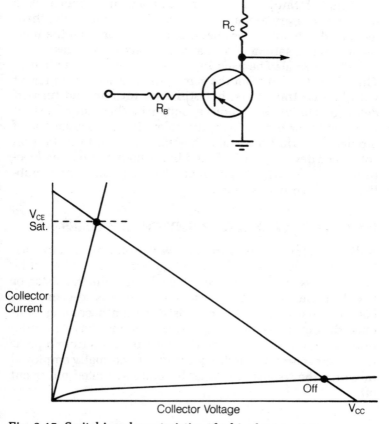

Fig. 2-15. Switching characteristics of a bipolar transistor.

Simplified design parameters for such a switching circuit are:

Base current	$I_B = V_{CC}/R_B$
Collector current	$I_C = V_{CC}/R_C$
	$I_C = h_{FR} \times I_B$
Bias resistor	$R_B = h_{fe} \times R_C$

where h_{fe} is the current gain of the transistor in the saturated mode. R_C is the load resistance in the collector line.

These formulas are all approximate. In practice, it is usually necessary to make the value of R_B about $1/4$ the theoretical value to allow for tolerances and ensure that the transistor remains saturated over a range of input voltages.

FETs can also be used in a similar manner as switches. They do not suffer from the same propagation delay present with bipolar transistors, but still have turn-on and turn-off delays due to interelectrode capacitance. These are of a similar order to, or higher than, the delay times characteristic of bipolar junction transistors. A great deal of information on solid state devices may be found in the book *Solid State Electronics Theory with Experiments* by M. J. Sanfilippo, published by TAB Books, Inc.

IMPROVING TRANSISTOR SWITCH-OFF TIMES

A direct method of reducing the switch-off time of a transistor is to reverse bias the base, but any such bias must not be allowed to exceed the reverse voltage limit of the transistor, or it will be damaged. An alternative method is to clamp the base voltage to prevent the transistor from becoming saturated during the switch-on period. This, too, has its limitations so when fast switching times are required from bipolar transistors, current switching circuits are normally employed in which the transistor neither becomes saturated nor is cut off.

DIODE SWITCHING

Diode switching characteristics are illustrated in simplified form in FIG. 2-16. When reverse biased there is only a very small leakage current. Application of forward voltage results in an immediate step to +V (forward conduction). The next application of reversed (-V) voltage, however, produces a transient due to stored charge effect, which then decays to the leakage current value. The peak transient reverse current can approach -V/R as a maximum, where R is the resistance

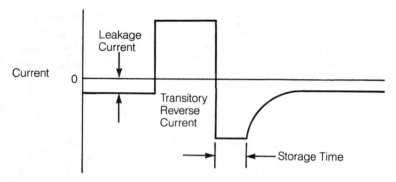

Fig. 2-16. *Switching characteristics of a diode.*

in the circuit. The time to reverse this charge, or storage time, varies with the type of diode and construction. In the case of ordinary diodes it can be a matter of milliseconds, reducing to nanoseconds in the case of high-speed switching diodes.

SCHOTTKY DIODES

The *Schottky diode* differs from conventional diodes in having a metal-to-semiconductor function at which rectification occurs. It has specific advantages over conventional junction diodes in that it does not exhibit carrier charge storage effects, thus enabling much faster switching speeds to be achieved. The voltage drop of the Schottky diode is also much less than that of an ordinary diode for the same forward current.

Diodes are commonly used as a clamp between the base and emitter of a transistor to prevent the transistor from entering saturation and to minimize propagation-delay time. It is readily possible to combine a Schottky clamping diode with a transistor as an integral device. Such a combination is called a *Schottky transistor*.

UNIJUNCTION TRANSISTORS

Unijunction transistors have two base contacts and an emitter. They become conductive (switch on) at a particular firing voltage, which typically ranges from 0.5 to 0.85 of the supply voltage. A particular application of the unijunction transistor as a switching device is to generate short pulses when supplied with a varied supply voltage, with pulse rates of up to 1 MHz readily obtainable.

THYRISTORS

An SCR is basically a silicon diode with an additional cathode electrode known as a *gate*. If the gate is biased to the same potential as the cathode, it does not conduct in either direction (except for a small leakage current). However, if the gate is biased to be more positive than the cathode, the SCR behaves as a normal diode; that is, it works as a switching element triggered by the application of a positive pulse to the gate.

The triac is similar in construction, except that it has both a cathode and anode gate; hence, it can be triggered by both positive and negative pulses.

SCRs and triacs are also known as thyristors. They are essentially alternating current switches, an SCR being triggered by the positive half of an ac voltage and a triac by both positive and negative halves of an ac voltage. Typical basic switching circuits are shown in FIG. 2-17.

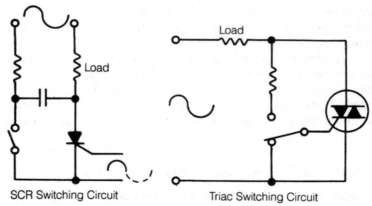

SCR Switching Circuit   Triac Switching Circuit

Fig. 2-17. *SCR and triac AC switches.*

BOUNCE-FREE SWITCHES

Mechanical switches commonly suffer from contact bounce when closed, which can give a spurious signal (especially when switching at rapid rates). This can be avoided by employing a *bounce-free* (or no bounce) *switch*. An example is shown in FIG. 2-18, employing an RS flip-flop as a follower for a mechanical switch. The effect of any contact bounce is now to raise both inputs to the flip-flop to logic 1, leaving the outputs unaffected.

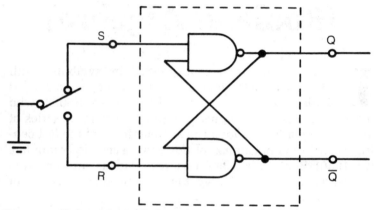

Fig. 2-18. Bounce-free switch.

3

Mathematical Logic (Boolean Algebra)

LOGIC functions can be expressed by symbols, truth tables, or mathematically. The latter is known as *Boolean algebra*, named after George Boole, who devised the system of representing logic through a series of algebraic equations as long ago as the middle of the last century. Until the appearance of the first electronic computers (in 1938) Boolean algebra was regarded as an academic mathematical exercise. Today it is a tool used by designers of logic circuits.

The basic symbols used in Boolean algebra are:

- • meaning a series condition or AND logic
- + meaning a parallel condition or OR logic
- − meaning negation or opposite condition or NOT logic

At this stage it is best to forget conventional arithmetic where • means multiply and + means add; otherwise, Boolean algebra may be confusing at first. Multiplication and addition do enter into working with Boolean equations, as explained later.

BASIC LOGIC

Basic logic symbols are shown again in FIG. 3-1 with equivalent equations in Boolean algebra.

Basic Logic 31

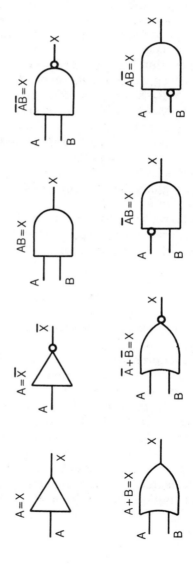

Fig. 3-1. Examples of annotated logic symbols with equivalent Boolean equations.

YES logic represents a simple continuous condition; that is, the output (X) is the same as the input (A). The corresponding mathematical equation is obviously $A = X$.

NOT logic represents a negation or opposite condition between input and output. Here the negation sign used in the mathematical equation becomes $\overline{A} = X$ (or $A = \overline{X}$).

AND logic requires that input A and B are both present before there is any output (a series condition), so the mathematical equation becomes $A \bullet B = X$.

NAND logic is the negation of AND, so here the equation becomes $\overline{A \bullet B} = X$. Alternatively, this equation can take the form $A \bullet B = \overline{X}$, which implies the same logic.

OR logic represents a parallel condition in that an input must be present at either A or B before there is an output. In this case, the + sign applies, and the mathematical equation becomes $A + B = X$.

NOR logic is the negation of OR, so the negation sign is added to give $\overline{A} + \overline{B} = X$.

The above basic equations are given for just two inputs. Exactly the same forms apply where there are more inputs. For example, the equation for an AND gate with five inputs is:

$$A \bullet B \bullet C \bullet D \bullet E = X$$

With the exception of NOT (which has only a single input and can only invert signals), each of the expressions for a logic function can be rearranged to obtain the others. This is a useful tool when designing logic circuits, for it enables the required functions to be rendered in the same logic dependent on the availability, or preferences for particular components, that is all in OR logic, all in AND logic, or all in NAND logic. This is done largely by using a NOT function (or single input NOR gate) as an inverter where necessary, and using the principle established by deMorgan's theorem which states that inversion changes the state of the logic each time it is applied; that is, from • to + or + to •.

For example, starting with the AND function:

$$A \bullet B = X$$

Inversion changes the AND ($\bullet$) to OR (+) logic:

$$\overline{A} + \overline{B} = \overline{X}$$

Inverting again gives a positive output:

$$\overline{\overline{A} + \overline{B}} = \overline{\overline{X}}$$

which is the same as:

$$A + B = X$$

In other words, double inversion has changed the function of an AND gate into OR logic working. All of these steps are shown in FIG. 3-2. At this stage, the basic rule to remember is that inversion changes the sign of the equation (except in a NOT gate) as well as changing the input. This is shown below:

Logic	Equation For Positive Output	With Inversion
OR	$A + B = X$	$\overline{A} \bullet \overline{B} = \overline{X}$
NOR	$\overline{A} \bullet \overline{B} = X$	$A + B = \overline{X}$
AND	$A \bullet B = X$	$\overline{A} + \overline{B} = \overline{X}$
NAND	$\overline{A} + \overline{B} = X$	$A \bullet B = \overline{X}$

OR Logic

Working with OR logic throughout, equations must use only the + sign, with inversion signals where necessary; that is, to change a $\bullet$ sign to a + sign producing the same function, and where necessary to give a positive output. The first example worked above shows how this is done with an AND

34 Mathematical Logic (Boolean Algebra)

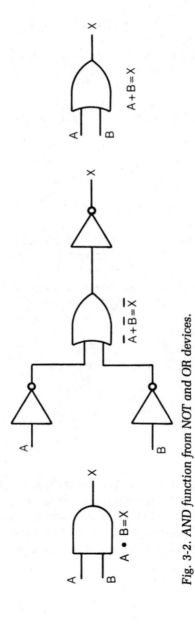

Fig. 3-2. AND function from NOT and OR devices.

gate. NAND and NOR functions can be obtained in a similar way.

The NAND function is already in OR logic:

$$\overline{A} + \overline{B} = X$$

Employing an OR gate to yield a NAND function theory requires inversion of both inputs as shown in FIG. 3-3.

The NOR function is in AND logic:

$$\overline{A} \cdot \overline{B} = X$$
$$\overline{\overline{A}} + \overline{\overline{B}} = \overline{X}$$

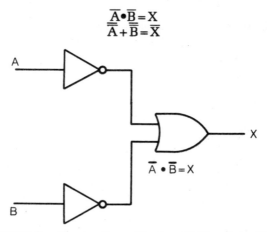

Fig. 3-3. *NAND function performed by two NOT and one OR device.*

Inversion on inversion puts the equation back to its original state, so this expression simplifies to:

$$A + B = \overline{X}$$

Thus the NOR function is performed in OR logic by an OR gate followed by a NOT gate for inversion. (See FIG. 3-4.)

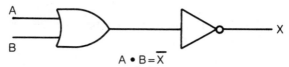

Fig. 3-4. *NOR function is performed by OR and NOT devices.*

AND Logic

Here the aim is to express all equations with the • (AND) sign. Obviously, an AND gate already does this; A•B=X and is shown in FIG. 3-5. Other logic functions can be determined from an AND gate as follows:

The OR function can be provided by inversion $\overline{\overline{A}+\overline{B}}=\overline{X}$
Check by inverting again $\overline{\overline{\overline{A}+\overline{B}}}=\overline{\overline{X}}$
Which is the same as $A+B=X$

The NOR function is already in AND logic ($\overline{A}\cdot\overline{B}=X$). The NAND function is devised simply by inversion of the output of an AND gate $A\cdot B=\overline{X}$.

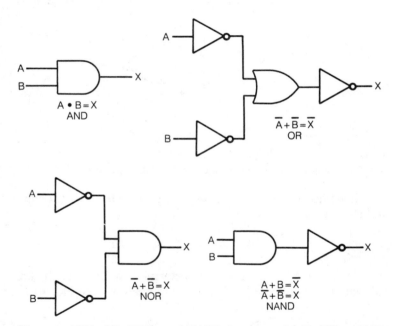

Fig. 3-5. AND, OR, NOR, and NAND functions devised from AND logic.

NAND Logic

Here, the requirement is to express all equations in the form (inverted AND).

To derive the OR function $A + B = X$
    then invert $\overline{A} \cdot \overline{B} = \overline{X}$
    invert again $\overline{\overline{A} + \overline{B}} = \overline{\overline{X}}$
    which is the same as $A + B = X$

To derive the NOR function $\overline{A} \cdot \overline{B} = X$
    invert $\overline{A} + \overline{B} = \overline{X}$
    invert for positive output $\overline{\overline{A} \cdot \overline{B}} = \overline{\overline{X}}$
    which is the same as $\overline{A} \cdot \overline{B} = X$

To derive the AND function $A \cdot B = X$
    invert $\overline{A} + \overline{B} = \overline{X}$
    invert for positive output $\overline{\overline{A} \cdot \overline{B}} = \overline{\overline{X}}$
    which is the same as $A \cdot B = X$

Derivations of the OR, NOR, and AND functions are shown in FIG. 3-6.

Exclusive OR

The OR gate, described previously, provides an output if one or more inputs has a value of 1. More specifically, it can be described as inclusive OR. There is a possible variation with a two-input OR gate where there is an output if one and only one of the inputs has a value of 1. This is known as the exclusive OR (it is also written XOR, sometimes described as non-equivalence) shown in FIG. 3-7. It has the truth table of TABLE 3-1. In other words, there is an output if $A = 1$ or $B = 1$, but not if the values $A = 1$, $B = 1$ occur simultaneously. The corresponding Boolean equation is:

$$(A + B)(\overline{A}\overline{B}) = X$$
$$\text{or } A\overline{B} + \overline{A}B = X$$

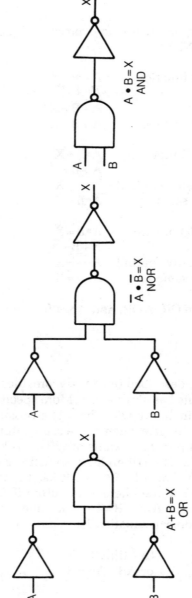

Fig. 3-6. OR, NOR, and AND functions devised from NAND logic.

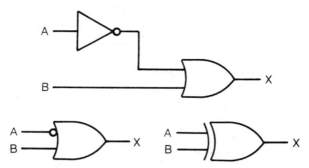

Fig. 3-7. Exclusive-OR shown in three different forms.

Table 3-1. Exclusive-OR Truth Table

Input		Output
A	B	X
0	0	0
0	1	1
1	0	1
1	1	0

Incidentally, notice in this equation that the period, or small point, between A and B has been eliminated. In fact, the point is almost never used and is understood as being there when two or more letters follow one another; that is, when they sit side by side.

A particular application of an exclusive OR is as a comparator or equality detector. For example, if the two input signals applied to the gate differ, there is an output. In this case the gate, in a sense, compares the two signals and detects the difference. Conversely, if the two input signals are identical, the exclusive feature means that there is no output. This absence of output indicates an equality of inputs.

Enable

Enable is an inhibit, such as provided by a NOT applied to one input of an AND gate as shown in FIG. 3-8 for a two-input AND gate with inhibit. The third input is called the *strobe* (S) or *enable* input, giving the truth table of TABLE 3-2

40 Mathematical Logic (Boolean Algebra)

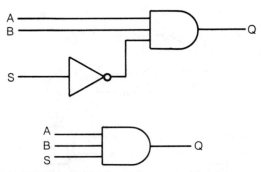

Fig. 3-8. ENABLE has an inhibit function on the S input.

Input			Output
A	B	S	Q
0	0	0	0
0	1	0	0
1	0	0	0
1	1	0	1
0	0	1	0
0	1	1	0
1	0	1	0
1	1	1	0

Table 3-2. ENABLE Truth Table

where the output is designated as Q. When a high is placed on the strobe input, a low is applied to the AND gate. This causes the AND gate to produce a low at its output. No matter what the other input sees, the output of the AND gate is always low.

As you can see from TABLE 3-2, there is an output, 1, only when A=1 and B=1, and S=0. The presence of an inhibit signal ($S=1$) holds the output at 0 irrespective of any possible combinations of A and B, even when $A=1$, $B=1$. The corresponding Boolean equation is:

$$AB\overline{S}=Q$$

SOLVING PROBLEMS

The basic process of designing logic circuits to meet particular requirements is to break down the problem into elementary yes-or-no or stop-go steps involving formal logic, and co-relating these steps as necessary. This means dealing with original truths (the facts of the question) called *propositions* and putting these together to arrive at an answer, or *syllogism*, based on the presence of these truths. Specifically, for example, if a single truth can be dealt with by NOT logic, the output responding to an input is either NOT (not true) or NOT NOT (true). Normally, however, more than one input is involved and there is some interrelationship between inputs, calling for the use of connections expressing the relationships. The most important of these are the AND and OR functions.

The following is a problem involving several propositions and connections, representing the prerequisites necessary to qualify for an executive position:

A. College Degree
OR B. Technical college with relevant certificates
C. At least 5 years of experience in a certain profession
D. Over twenty-five years of age
E. Not married

In plain language the basic relationship is:

A OR B AND C AND D AND NOT E

The corresponding Boolean equation is:

$$(A+B)CD\overline{E}=X$$

42 Mathematical Logic (Boolean Algebra)

An immediate solution employing AND, OR, and NOT logic gates is shown in FIG. 3-9. This also follows directly from the Boolean equation. Suppose, however, that only AND and NOT devices are available. This means that the problem must be solved in AND logic only. This can be started by inverting the original equation thus:

$$(\overline{A}\overline{B}) + \overline{C} + \overline{D} + \overline{\overline{E}} = \overline{X}$$

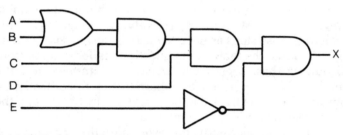

Fig. 3-9. Problem solution using AND, OR, and NOT gates.

Now invert again:

$$\overline{(\overline{A}\overline{B})}\,\overline{\overline{C}}\,\overline{\overline{D}}\overline{E} = \overline{\overline{X}}$$

Note here that by containing $(\overline{AB})$ as one term in a bracket it does not change its state on inversion. Now remove double inversions as they merely mean using pairs of NOT devices to get back to the original output:

$$\overline{(\overline{A}\overline{B})}CD\overline{E} = X$$

The bracketed term $\overline{(\overline{A}\overline{B})}$ remains something of a problem as it still contains double inversion. However, since we are restricted to NOT and AND devices this is really no problem at all, as it can be accommodated by a NOT device in each input to an AND, and a further NOT in the output. The final circuit in AND logic is shown in FIG. 3-10.

Given no restrictions on availability of components, further solutions can be worked in Boolean algebra to see if any

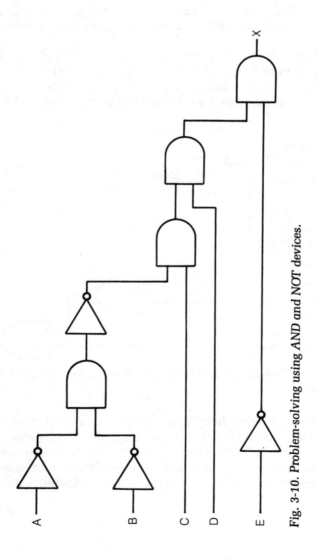

Fig. 3-10. Problem-solving using AND and NOT devices.

simpler circuit can be derived. There is, in fact, using NOR logic $(\overline{+})$:

Starting with	$(A+B)CD\overline{E}=X$
and inverting	$\overline{(A+B)}+\overline{C}+\overline{D}+\overline{\overline{E}}=\overline{X}$
inverting again as a whole	$\overline{\overline{(A+B)}+\overline{C}+\overline{D}+\overline{\overline{E}}}=\overline{\overline{X}}$
and removing double inversions	$\overline{\overline{(A+B)}+\overline{C}+\overline{D}+E}=X$

Remembering that bracketed inputs, $\overline{(A+B)}$ in this example, must be directed to one separate (NOR) device, the final circuit then works out as in FIG. 3-11. This saves two components compared with the AND logic circuit of FIG. 3-10.

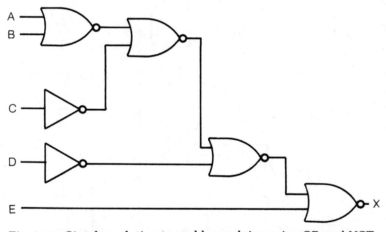

Fig. 3-11. Simpler solution to problem solving using OR and NOT devices.

Whether solution by Boolean algebra is quicker or simpler than design by digital logic diagrams is debatable. For some people it is, for others it is not. Where it does have a definite advantage is in positive elimination of unnecessary components by making it simple to spot and remove double inversions.

BOOLEAN ALGEBRAIC THEOREMS

Most problems can be solved by applying the appropriate Boolean algebra theorems, the basic rules under which Boolean algebra works. Only one has been mentioned so far, deMorgan's theorem, which is:

$$\overline{ABC} = \overline{A} + \overline{B} + \overline{C}$$
$$\text{or } \overline{A + B + C} = \overline{A}\overline{B}\overline{C}$$

There are numerous others, some obvious, others rather more difficult to understand at first. Those which may be of particular significance are:

- $\overline{\overline{A}} = A$ or $\overline{\overline{B}} = B$, etc. Double inversion returns the function to its original form.
- $AA = A$. This means that with an AND device, application of the same signal to both inputs will result in the same output.
- $A + A = A$. The same as above, but in this case relating to an OR device.
- $A\overline{A} = 0$ or $A\overline{B} = 0$. With one input inverted, there is no output from an AND device.
- $A + \overline{A} = 1$ or $A + \overline{B} = 1$. With one input inverted, provided one has a value of 1, there is always an output from an OR device.

A number 1 appearing in a Boolean equation means that one signal is always applied, while a 0 means that there is no signal at that particular input. (The numbers 1 or 0 in this case replace A or B, etc., on a particular input diagram.) Therefore the next equations can be stated as follows:

- $A0 = 0$ (the AND function can never be completed with one input always at 0).
- $A1 = A$ (the AND function is completed with a single input A when the second input is 1; the output is governed by the value of A).

46 Mathematical Logic (Boolean Algebra)

- $A + 0 = 0$ (the OR function is complete with one input signal if the other input signal is 0).
- $A + 1 = 1$ (the OR function is complete with a single input when the second input is 1. Compare this with the AND equivalent).

Functions enclosed by a bracket are subject to normal algebraic treatment when expanded, as shown next:

- $A(B+C)$ or A AND (B OR C) becomes $AB + AC$ (A AND B OR A AND C)
- $(A+B)(C+D)$ OR (A OR B) AND (C OR D) becomes $AC + AD + BC + BD$ (A AND C OR A AND D OR B AND C OR B AND D)

Checking by writing out in words and comparing with the original expression verifies if the original expansion is correct or not; i.e., $A + (AB) = A$. This is self explanatory on spelling it out; A OR A AND B. It is an OR function satisfied if only A is present.

In the example $A + (\overline{A}B) = A + B$, this is an OR function, so it is satisfied if A OR B is present. This can be shown with the following equation:

$$A \text{ OR NOT } A \text{ AND } B = A \text{ OR } B$$

4

Logic Circuit Devices

BASICALLY all the functions of a logic switching system can be provided by NAND/NOR gates, or by either an AND or OR gate(s) and inverters. The former is the preferred method since AND/OR circuitry has a number of practical limitations. If AND/OR elements are cascaded, for example, each produces some attenuation of the signal which may require additional amplification at certain stages, thereby complicating circuit design. With NAND/NOR circuit design, this is not necessary since the main requirement here is in observing the maximum number of inputs (fan-in) and outputs (fan-out) provided by each element.

Initially, all electronic logic circuits were constructed from discrete components such as transistors and diodes for active elements and resistors and capacitors for passive elements. Typically, these yielded printed circuit modules about 1 inch to 2 inch by 1 inch for assembly into complete circuits. These have now been almost entirely replaced by integrated circuits (ICs) offering the performance capabilities of numerous interconnected modules in a single miniaturized package. Besides offering very great reductions in weight and size, as mentioned in chapter 1, integrated logic circuits also have the advantages of greater reliability and greater speed of operation. They are now generally cheaper than all the components needed to construct discrete modules covering the same functions.

Some integrated circuits have the disadvantage of lower signal levels in the order of 0.8 to 2 volts as compared with 6 to 12 volts (or even 24 volts) normally employed with discrete modules. This renders the IC more susceptible to noise and can place a premium on component location, lead length, and grounding requirements. However, the widely used CMOS ICs can be used over a wider voltage range and have very high noise immunity.

As with discrete component modules, IC logic circuits are based on the same components of transistors, diodes, etc., although in very much miniaturized form. Schematically, therefore, the two forms of circuits are identical, although for the purpose of use only the external connection points of the IC normally need to be identified.

The diode-resistor network shown in FIG. 4-1 provides positive AND logic. With all inputs A, B, C...N positive (logic 1), all the diodes are reverse biased and do not conduct, giving an output of +V (logic 1). In the absence of any

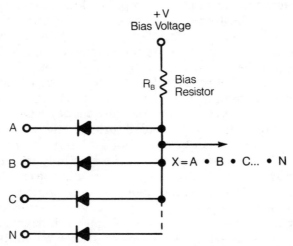

Fig. 4-1. A diode-resistor logic network using positive AND logic and negative OR logic.

one input, that diode conducts, causing the output to fall to 0.

The same circuit with negative logic (−V corresponding to logic 1) works as an OR gate giving a 1 output in the presence of any input. Equally, if the bias voltage is made more positive than logic 1, all diodes conduct when all the inputs are present together, clamping the output to logic level 1.

The network shown in FIG. 4-2 has the diodes connected in the opposite manner to those of FIG. 4-1. This time, with positive logic (+V as input) it works as an OR gate and with negative logic (−V as input) it works as an AND gate. Again there is the possibility of clamping the output if required.

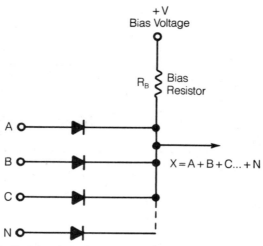

Fig. 4-2. A diode-resistor logic network using positive OR logic and negative AND logic.

The disadvantage of these networks is that if the circuits are cascaded, the input current to any one circuit must be provided by the circuit preceding it. This means that relatively low values of bias resistors must be used in order to maintain the required drive currents. In practice this may not be possible and buffer amplifiers have to be inserted between stages.

DIODE-TRANSISTOR LOGIC (DTL)

Diode-transistor logic overcomes the limitation of cascading by incorporating a transistor amplifier in the output circuit. A typical positive logic NOR gate of this type is shown in FIG. 4-3. Here any input going positive (logic 1) causes the base of the transistor to go positive with respect to the emitter and cut off. The output is then logic 0 (no current flow through the collector circuit). When all of the inputs are logic 0, the base of the transistor is negative, yielding a collector output approaching the emitter value or logic 1. Worked with negative logic ($-V$ = logic 1), this circuit provides a NAND function.

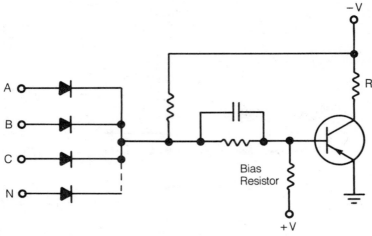

Fig. 4-3. Diode-transistor logic (DTL).

DTL logic was originally widely produced in IC form operating at speeds of 2-20 MHz with logic levels between 0.5 and 5 volts and for power supplies between 3 and 6 volts. It has now been replaced by simpler and more efficient networks such as transistor-transistor logic (TTL) and more exotic devices.

RESISTOR-TRANSISTOR LOGIC (RTL)

Resistor-transistor logic is another network form which was widely used for discrete modules, but found less suitable in IC form because of its low logic levels (about 1 volt) on 3-4 volt supplies. It also has poor fan-out (limited number of outputs) and noise immunity. It is still of interest for discrete module construction since it is a simple and straightforward circuit with a wide tolerance for variations in component working values (FIG. 4-4).

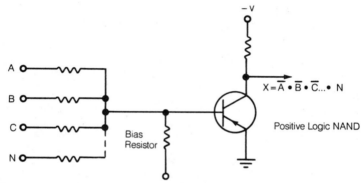

Fig. 4-4. Resistor-transistor logic (RTL).

DIRECT-COUPLED-TRANSISTOR LOGIC (DCTL)

With DCTL logic, only transistors are used as the switching elements with the advantage of requiring only one low voltage supply with low power consumption and fast switching speeds. It is attractive for producing IC NAND and NOR gates utilizing a minimum of components. (See FIG. 4-5.)

Disadvantages of this network are that each input requires its own transistor, and these transistors must have uniform characteristics, making DCTL an unattractive choice for construction of discrete modules. These limitations are not so significant in IC construction, but this type of IC circuit is still relatively susceptible to noise.

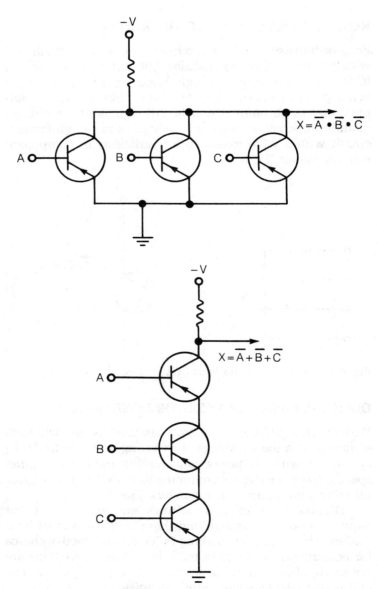

Fig. 4-5. Direct-coupled-transistor logic (DCTL).

EMITTER-COUPLED-TRANSISTOR LOGIC (ECTL)

In the emitter-coupled-transistor logic the transistors are not allowed to saturate fully and switch a constant current from one transistor to another. For this reason it is sometimes called Current-Mode Logic (CML). It is considerably less susceptible to noise than DCTL and has much higher switching speeds.

FIGURE 4-6 shows the network for a NAND gate. Here the bias voltage maintains a constant current through T_4 if all the inputs are at a positive level (0V = logic 1). The output at X_1 is then negative or logic 0 ($\overline{A}\overline{B}\overline{C}$). It is therefore positive at X_2 or logic 1 (ABC). If any input goes negative ($-V$ or logic 0) its transistor will conduct through R_e causing T_4 to cut off. In this case output 1 goes to ground (logic 1) and output 2 goes to $-V$ (logic 0). A feature of this circuit is that it provides a NAND function at output 1 and an AND function at output 2.

TRANSISTOR-TRANSISTOR LOGIC (TTL)

In transistor-transistor logic transistors are connected in the common-base mode; a typical circuit is shown in FIG. 4-7. All NOR inputs have to be negative (logic 0) for the output to go positive (logic 1). Any input going positive causes its transistor to conduct and transistor T_4 to cut off. Hence, the output is then 0. Rendered in IC form, a multiple-emitter transistor is normally used with the corresponding circuit shown in FIG. 4-8.

Circuits of this type are fast switching (4 to 50 MHz) with good noise immunity, and are relatively simple to produce. They are one of the main types used in digital ICs. A typical TTL device can drive up to ten TTL inputs (has a fan-out of ten), but should not be connected with outputs of different families in parallel unless having a modified output stage.

Most of the range of IC devices in TTL are also produced in low-power Schottky logic based on Schottky diodes and Schottky transistors. These have the advantage of faster

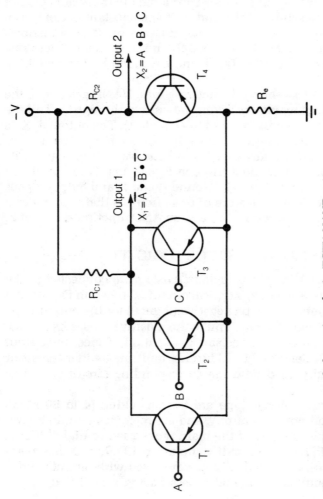

Fig. 4-6. Emitter-coupled-transistor logic (ECTL) NAND gate.

Transistor-Transistor Logic (TTL) 55

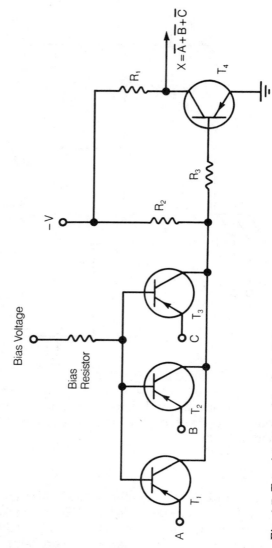

Fig. 4-7. Transistor-transistor logic (TTL).

56 Logic Circuit Devices

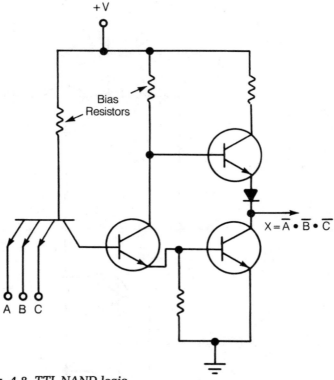

Fig. 4-8. TTL NAND logic.

switching speeds and lower current consumption (only about 25 percent of the operating level of typical TTL devices).

MOSFETS

The metal-oxide-semiconductor field effect transistor (MOSFET) is basically a special form of FET often just called MOS. It has the attraction of being particularly suitable for extreme miniaturization allowing large and very large scale integration (LSI and VLSI). MOS devices are thus widely used in digital electronics as logic gates, registers, and memory arrays. MOSFET circuits consist entirely of FETs (except for parasitic capacitors in certain dynamic applications), but can be made with a zener diode between the gate and substrate of each, or selected, FET(s). The object of this is to protect the gate from excessive voltages. Under normal operation, the zener diode remains open with no effect in the circuit, but the maximum gate voltage that can arise is limited to the maximum value of the zener voltage. Examples of MOSFET gate circuits are shown in FIG. 4-9 together with standard circuit symbols for a MOSFET. Variations on the symbols used for MOSFETs are shown in FIG. 4-10.

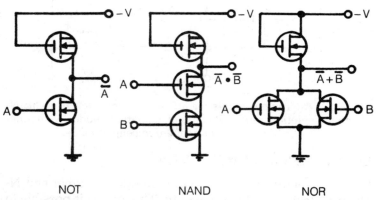

      NOT   NAND   NOR

Fig. 4-9. *Typical MOSFET gate circuits.*

MOSFET gates are, in fact, examples of direct-coupled-transistor logic (DCTL). The only basic difference is that because of the high density of components on the same chip it becomes necessary to minimize power consumption in large scale integration, although their efficiency, in terms of power performance, is superior to that of ordinary bipolar DCTL gates.

There are subtle differences between the characteristics of MOSFETs and FETs. The drain resistance of a MOSFET is lower than that of an FET, while the resistance between gate and drain or gate and source is higher. In all cases, however, these resistances are extremely high and virtually equivalent to open circuits when shunted by external circuit resistors.

58 Logic Circuit Devices

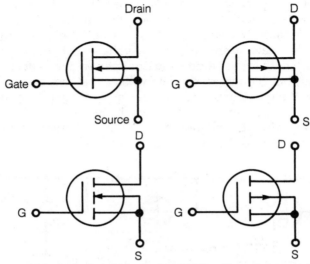

Fig. 4-10. Alternate symbols for MOSFETs. From left to right: N-channel depletion mode, P-channel depletion mode, N-channel enhancement mode, and P-channel enhancement mode.

COMPLEMENTARY MOS (CMOS)

Complementary MOS or CMOS employs P-channel and N-channel devices on the same chip. This makes it possible to reduce power dissipation to very low levels as small as 50 nanowatts. Like MOSFETs, the basic CMOS device is an inverter. Combinations of these devices can be used to provide CMOS NAND and NOR gates. About the only disadvantages shown by MOSFET and CMOS devices are their slower speed of working as compared to some other devices, and certain high frequency limitations inherent with field effect transistors due to internal capacitance effects.

MOS LOGIC

MOS logic elements are now widely used and have largely taken over from TTL for integrated circuits. The extremely high component density possible means that large memories, shift registers, and circuits of this type, can be produced in very compact packages. While functions performed are basically similar to those of other logic devices, the behavior and specific characteristics of MOS and related devices do differ appreciably and need to be appreciated.

The working mode of asynchronous MOS circuits is similar to that of other transistor gates using FETs. This differs from bipolar junction transistors in that MOS devices are unipolar, have a high input resistance, and are generally less noisy than bipolar transistors. Their main disadvantage is the lower gain and the susceptibility of the thin silicon layer of the gate to damage by excessive voltage. MOSFETs are also slower than bipolar transistors.

The majority of such circuits use P-channel enhancement mode MOS devices, where the drain supply is a negative potential and thus they work with negative logic. In other words, a high negative voltage represents a logic 1. The supply voltage for such devices commonly ranges from −10 volts to −20 volts, with logic 1 having a value on the order of −10 volts. With higher voltages, logic 0 normally lies at a level of −2.5 to −5 volts.

P-channel and N-channel MOSFETs can also be used in complementary configuration to operate with positive logic. The particular advantage of this is that N-channel devices are faster, and so such circuits can have faster switching times than P-channel devices. Two basic complementary MOS gates are shown in FIG. 4-11. Other types of MOS devices include low threshold PMOS, VMOS, DMOS, and HMOS.

PMOS devices incorporate silicon gates in place of input and output FETs to allow easier interfacing to TTL and to increase switching speed of the device. Additionally, switching speeds of three times that of NMOS or PMOS (N-channel or P-channel, respectively) devices is achieved in VMOS devices by reduced gate resistance, a result of a V shape cut in the gate region. DMOS, or double-diffused doping MOS devices, dissipate only about one half the power of standard MOS but at the sacrifice of switching speed. In HMOS, or high-performance MOS devices, switching speed is extremely fast and power consumption is minimal, but this type of device has been prohibitive due to excessive cost.

60 Logic Circuit Devices

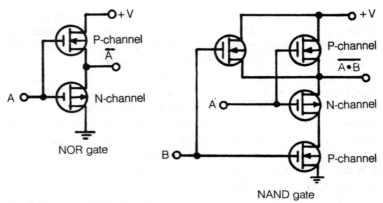

Fig. 4-11. Two basic CMOS gates.

CLOCKED MOS CIRCUITS

MOS circuits are particularly suitable for *synchronous*, or *clocked*, systems. These are generally referred to as dynamic MOS circuits. The advantage here is that average power consumed by the system is reduced. However, where gates are cascaded it is necessary to have more than one pulsed supply to allow for the time it takes the output voltage to reach a steady state. These pulsed voltages are then applied sequentially to the system, giving two-phase systems, three-phase systems, four-phase systems, and so on.

DYNAMIC MOS INVERTERS

A basic circuit for a dynamic MOS inverter is shown in FIG. 4-12, operating as mentioned before with negative logic. This requires a train of pulses to operate. At logic state 0 (no pulse) both transistors are switched off and there is minimal power consumed. With the appearance of a negative pulse, both transistors are switched on and conduct with output being the inversion of the input. If $A = 1$ then $Q = 0$, or if $A = 0$ then $Q = 1$. The output is held on for the duration of the pulse by the charge on the output capacitor C.

A particularly important feature of a dynamic MOS circuit is that the parasitic capacitance between gate and sub-

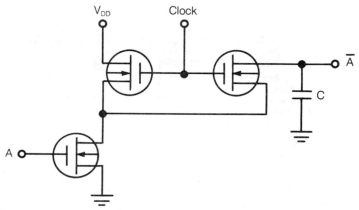

Fig. 4-12. Dynamic MOS inverter.

strate inherent in a MOSFET is used to provide temporary memory or storage capacity with a time constant on the order of milliseconds. This storage can be refreshed and made permanent by the application of a clock waveform of suitable frequency, such as giving pulse times substantially longer than the time constant of delay. A typical refresher frequency is normally 1 kHz or longer.

DYNAMIC MOS NAND GATES

A basic circuit for a dynamic MOS NAND gate is shown in FIG. 4-13. This is similar to the static NAND gate of FIG. 4-9 except for the additional FET which works as a switching element controlled (switched on and off, respectively) by the clock pulse. Again, in the off condition all transistors are off and power dissipation is minimal.

The dynamic MOS NOR gate is similar to a static NOR gate with an additional FET acting as a switch for the clock pulse, shown in FIG. 4-14.

HANDLING MOS DEVICES

MOS integrated circuits are more readily damaged than other devices and thus need handling and mounting with care.

62 Logic Circuit Devices

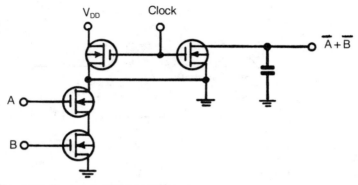

Fig. 4-13. Dynamic MOS NAND gate.

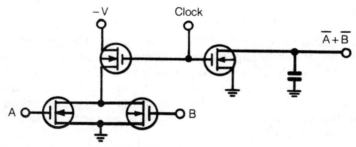

Fig. 4-14. Dynamic MOS NOR gate.

They are easily damaged by static charges or transient high voltages. Ideally, they should be handled on a conductive surface such as a metal tabletop, to which the person handling the device is also connected by a metal bracelet, or conductive cord or chain.

Similar recommendations apply when mounting MOS devices on a printed circuit board. If it is impractical to ground the printed circuit board, then the person mounting the circuits should touch the board first to discharge any static before the MOS device is brought into contact with the board. In practice, the most modern CMOS ICs are difficult to damage and the only precaution necessary is to store the chips in conductive plastic carriers, so that all pins are shorted together.

INTEGRATED CIRCUITS AND MINIMIZATION

The ready availability of complex circuitry in integrated circuit chips has considerably changed attitudes towards circuit design and construction. Medium scale integration (MSI) can offer dozens of gates in a single package; large scale integration (LSI) hundreds of gates in a single chip; very large scale integration (VLSI) thousands of gates in a single chip. These chips are used to build other, more complex digital logic circuits. The question of minimization or the elimination of redundant gates then becomes relatively unimportant. A standard IC package for a computer, decoder, shift register, read-only-memory, etc., may provide more internal circuits than are actually required, but still offers the most straightforward, and cheapest, solution even if all of the pins are not used.

This has influenced design technique too. Instead of designing a specific, individual circuit as in the days of module construction with discrete components, the circuit designer is more and more having to accept what is predesigned in an IC package and use the facilities it provides accordingly. This means the designer has to work with subsystems, rather than specific gates or other binary units. This has resulted in new design techniques being developed for implementing circuit performance requirements with IC subsystems.

STANDARD IC GATES

Integrated circuits are produced in a variety of packages. The most common are the TO5 (Transistor Outline) style can, similar in size and appearance to a transistor, but with as many as 12 leads emerging from the bottom; and the flat package. The latter is of rectangular wafer form or in a dual-in-line package (DIP) with connections brought out at right angles from both sides (FIG. 4-15). The DIP IC is larger, much easier to mount on printed circuit boards, and also cheaper to produce.

Common forms of digital IC gates are quadruple two-input NAND, triple three-input NAND, dual four-input

64 Logic Circuit Devices

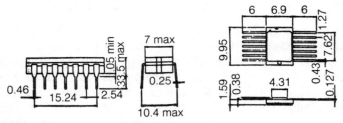

Fig. 4-15. Dual-in-line and flat IC packages.

NAND, single eight-input NAND, quadruple two-input NOR, quadruple two-input AND, inverters, and buffers, but there are many more. Such gate circuits are available in most logic families, particularly DTL, TTL, DCTL, and ECTL. The limitation on the number of gates per chip is normally set by the number of pins available. As an example, common numbers for a flat package are 14, 16, 24, 28, and 40 leads.

Where two different families of ICs may be involved in a complete circuit (such as TTL and MOSFET) the question of compatibility can arise because of the difference in operating voltage levels. Such differences can be accommodated by buffer circuits dropping a higher level voltage to a lower level voltage where required. These types of ICs are referred to as *level translators*.

MULTIPLE GATE ICS

Integrated circuits commonly contain multiple circuits or complete subsystems in a single package such as dual, triple, and quadruple gates; hex buffers and inverters; flip-flops and latches; shift registers; counters; multiplexers; mnemonics; display drivers; and arithmetical circuits. All such packages may appear similar except for the number of leads. The designation of the leads is therefore of primary importance.

Pin numbering reads around the IC left to right then right to left, as shown in FIG. 4-16. Note also that some ICs do not have a notch marking the pin 1 position, but a dot mark instead. For example, FIG. 4-17 shows a family of NOR gates with the internal devices shown in symbolic form together

Multiple Gate ICs 65

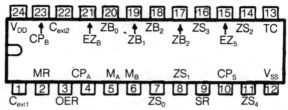

Fig. 4-16. Conventional method of lead or pin-out numbering. This IC is a real time S-decoder counter.

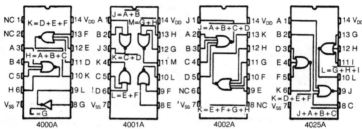

Fig. 4-17. Family of IC NOR gates.

with their connection to external leads. Externally there is no difference in the appearance of these packages, although they have quite different functions and external connections. The 400A is a dual three-input NOR gate (two gates plus inverter). The 4001A is a quad two-input NOR gate (four gates). The 4002A is a dual four-input NOR gate (two gates). The 4025A is a triple three-input NOR gate (three gates). A, B, C, D, etc. are inputs to the gates while J, K, L, etc. are gate outputs. Additionally, V_{DD} and V_{SS} are the supply voltages, with V_{SS} being the most negative power supply to the device. Finally, N.C. simply means no connection. This provides all of the information necessary to connect the chosen IC into a given circuit.

If the circuit is to be designed around the IC, then the electrical characteristics as specified by the manufacturer need to be known as well. A logic diagram of the IC can also be helpful. FIGURE 4-18, for example, is a logic diagram for an 8-input NOR gate IC (HEF 4078B).

66 Logic Circuit Devices

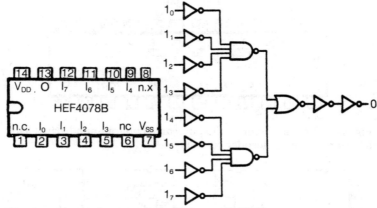

Fig. 4-18. Logic diagram for an 8-input IC NOR gate.

IC BUFFERS

Individual buffer circuits are produced in IC form, the usual number being six contained in a standard 16-pin package. These may be inverting buffers or non-inverting buffers, described as hex inverting buffers or hex non-inverting buffers as shown in FIGS. 4-19 and 4-20.

Where buffers are provided with input protection as shown in FIG. 4-21, input voltages in excess of the noted supply voltage for the buffers can be accepted. Such buffers can also be used to convert logic levels of up to 15 volts to standard TTL levels.

Hex buffers are also produced with three-state outputs as illustrated in FIG. 4-22. Here the three-state outputs are controlled by two enable inputs. A predetermined number of buffers can then be made to assume an off state via the appropriate enable signal regardless of the input conditions.

SCHMITT TRIGGER

The Schmitt trigger is another hex (six gate) IC form, this time in 14-pin packages. These trigger circuits are available in inverting and non-inverting forms as shown in FIG. 4-23.

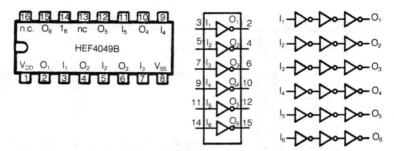

Fig. 4-19. Hex inverting buffer IC.

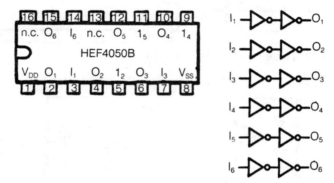

Fig. 4-20. Hex non-inverting buffer IC.

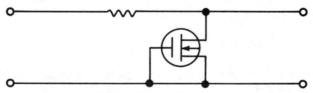

Fig. 4-21. Input protection for buffer circuits.

COMPLEX ICS

Integrated circuits embodying complete subsystems may have 14, 16, 24, or even 40 leads, each lead specifically designated. This may be in words and/or code letters. Abbrevia-

68 Logic Circuit Devices

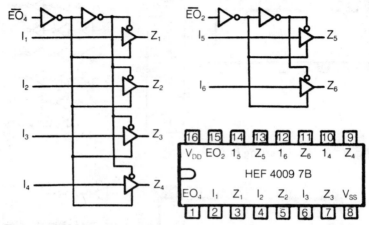

Fig. 4-22. Three-state hex non-inverting buffer.

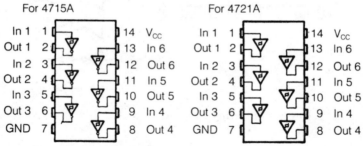

Fig. 4-23. Hex Schmitt trigger IC.

tions commonly used are:

- A0, A1, A2, etc., for inputs (especially address inputs)
- 01, 02, 03, or Q0, Q1, Q2, etc., for outputs
- D for data input
- E for enable
- El for latch enable
- C, Ck, Cp for clock (input)
- $\overline{CE}$ for clock enable
- R for reset

- S0, S1, S2, etc., for select inputs
- ST for strobe input
- Cl or CL for clear
- R/W for read/write input

DIGITAL FAMILIES COMPARED

DTL, originally used for the production of NAND gates, is now largely regarded as obsolete for IC production. Its chief limitations are that it has limited fan-out and a relatively high propagation delay (typically 30 ns per gate). Only a low voltage supply is necessary, however, and power dissipation is low.

TTL has similar or slightly higher power dissipation, but smaller propagation delay and very good noise immunity. MOS and CMOS devices are slower than TTL and also more sensitive to capacitance loading.

CMOS is particularly suited to LSI and VLSI because of the very small device size possible and the higher potential packing density. TTL elements are generally produced in SSI and MSI complexity.

In terms of power dissipation, low power Schottky (LS-TTL) and TTL are similar, with MOS lower and CMOS substantially lower. Some comparative data is summarized in TABLE 4-1.

Table 4-1. Comparisons of Digital Families

	DTL	Standard TTL	MOS	Low power Schottky	CMOS
Supply voltage	low	low	15–20	-	1.5–18
Power dissipation per gate mW	8–12	12–22	0.2–10		0.1–1
Quiescent power mW		10			
Propagation delay ns	30	10	300	10	70
Clock frequency MHz	8	35	2	45	5
Noise immunity	fair to good	very good	low	very good	very good
Fan out	8	10	20	10	50

5

Flip-Flops and Memories

THERE are a number of different types of logic devices, or more appropriately, combinations of logic devices that perform the function of storing a *binary digit* (bit). One of these is the flip-flop. It is made up of several gates so arranged that placing a 1 or 0 on its input can cause it to hold (memorize) that 1 or 0 even when the input is removed. The flip-flop latches on to the input level and is therefore also called a latch. Dfferent kinds of flip-flops are used for various purposes; some are discussed in this chapter.

RS FLIP-FLOPS

The RS flip-flop can be configured in different ways, always with the same expected results, using digital logic gates. The first of these is the RS NOR latch shown in FIG. 5-1. This latch uses two cross-coupled NOR gates to perform the latch function. *Cross-coupled* means that the output of NOR gate 1 acts as one of the inputs to NOR gate 2 while the output of NOR gate 2 acts as one of the inputs to NOR gate 1. The truth table is listed in TABLE 5-1. Notice that the RS NOR latch has two outputs, Q and $\overline{Q}$ (not Q). $\overline{Q}$ will always be the opposite of Q. If Q is 1 then $\overline{Q}$ is 0, and if Q is 0 then $\overline{Q}$ is 1.

72 Flip-Flops and Memories

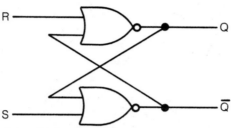

Fig. 5-1. The RS NOR latch (flip-flop).

R	S	Q	Q̄	Mode
0	0	NC	NC	Hold
0	1	1	0	Set
1	0	0	1	Reset
1	1	0	0	Disallowed

Table 5-1. RS NOR Latch Truth Table

The operation of the RS NOR latch is such that on power up of the latch you must assume that one of the gates will switch first and cause a condition of a 1 or 0 on the Q output. This then determines the operation of the flip-flop. If Q is 1 (high) then Q̄ is 0 (low). With a 1 on the S (set) input the Q output stays high and is said to be set. The Q output remains high even with both inputs removed. The only time Q goes low (flip) is when there is a 1 on the reset input and a 0 on the set input. To get Q to go high again (flop) the set input must be high with the reset line low.

The RS NAND latch is illustrated in FIG. 5-2. It performs a latch function also but because it uses NAND gates, the same

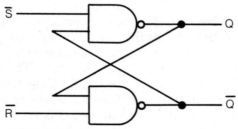

Fig. 5-2. The RS NAND latch.

input conditions produce opposite results. Its truth table is listed in TABLE 5-2.

The clocked RS flip-flop of FIG. 5-3 has two additional NAND gates that allow either the reset or set pulse to trigger the flip-flop, but only when the clock input is positive. Its truth table is listed in TABLE 5-3.

Table 5-2. RS NAND Latch Truth Table

$\overline{R}$	$\overline{S}$	Q	$\overline{Q}$	Mode
0	0	1	1	Disallowed
0	1	0	1	Reset
1	0	1	0	Set
1	1	NC	NC	Hold

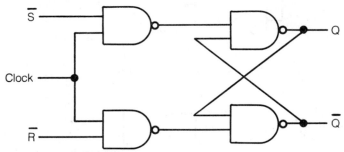

Fig. 5-3. A clocked RS flip-flop.

Table 5-3. Clocked RS Flip-Flop Truth Table

CK	R	S	Q	$\overline{Q}$	Mode
0	X	X	NC	NC	Disable
⊓	0	0	NC	NC	Hold
⊓	0	1	1	0	Set
⊓	1	0	0	1	Reset
⊓	1	1	0	0	Disallowed

D FLIP-FLOPS

D flip-flops are also called *data latches*. As long as the clock input is high, Q follows the value of input D. If D is high when the clock (Ck) is high, then the output (Q) is high. If D goes low while the clock is still high, then Q goes low. In other words, Q is at the same level as D as long as the clock is high. However, once the clock input goes low, the output at Q remains at whatever the last value of D was just prior to the clock going low. The flip-flop latches to the last value of D while the clock input was high. A D flip-flop is shown in FIG. 5-4. Its truth table is shown in TABLE 5-4.

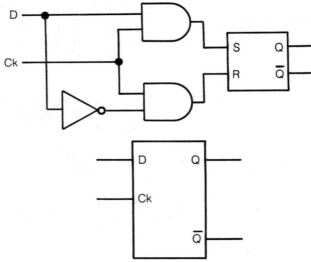

Fig. 5-4. *A D-type flip-flop can be constructed as shown on the top, but comes in a single IC and uses the symbol on the bottom.*

CK	D	Q
0	X	Last State
1	0	0
1	1	1

Table 5-4. *D Latch Truth Table*

To see how a D latch operates as a memory device, look at FIG. 5-5. Here you see four D latches with their clock lines tied together. This is the concept of temporary storage of a word (4 bits) of memory. When the clock input goes high, the

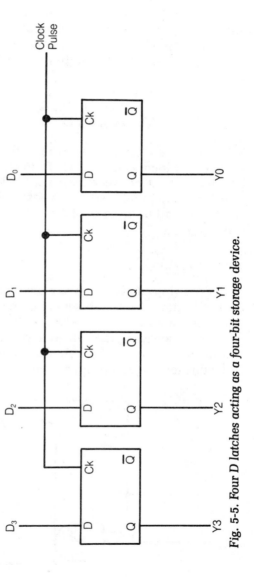

Fig. 5-5. Four D latches acting as a four-bit storage device.

input data is loaded into the flip-flops. The levels of this input data are also seen at the output. As soon as the clock goes low (it goes low on all the latches at the same time since they're tied together) the output retains this data. As an example:

$$\text{If } D_3, D_2, D_1, D_0 = 1011$$
$$\text{Then } Q_3, Q_2, Q_1, Q_0 = 1011$$

As soon as Ck goes low, 1011 is the output data that is retained. As long as Ck is low, D_3 to D_0 can change all day long, but Q_3 to Q_0 always remains 1011. This is a good example of what can be referred to as a *basic memory circuit*. Later you will see all of these D latches incorporated into a single IC package.

JK FLIP-FLOPS

The JK flip-flop can function as a clocked RS flip-flop or as a toggle flip-flop. It can also serve in a number of specialized functions. In addition, there are no forbidden (ambiguous) conditions, meaning that all four possibilities in its truth table are equally valid. A JK flip-flop is shown in FIG. 5-6. The truth table for the JK flip-flop is shown in TABLE 5-5. This is a

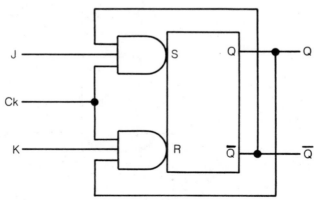

Fig. 5-6. *Logic diagram for a JK flip-flop.*

Table 5-5. JK Flip-Flop Truth Table

J	K	Q, (after Ck)
0	0	No Change
0	1	Resets
1	0	Sets
1	1	Toggles

positive edge triggered flip-flop meaning that clocking occurs when the clock goes from low to high (from 0 to 1).

THE JK MASTER-SLAVE FLIP-FLOP

The JK master-slave flip-flop is actually the end product of the previous flip-flops discussed so far. It eliminates timing problems associated with the simpler latches including a problem called *racing*. Racing occurs when a flip-flop toggles more than once during a positive clock edge. A logic diagram for the JK master-slave flip-flop is shown in FIG. 5-7. Notice that the clock input is provided directly to the master section and also to the slave section, but there, through an inverter. This low level clock into the slave section locks out any data input to that section.

With the arrival of a clock input to the master section, either J or K is ready to cause this section to change state. This ready state on the J or K input is a function of the outputs of the slave section. If the slave is in the reset state, the master can set, and if the slave is in the set state then the master can reset.

If it is assumed that the slave section is in the reset condition, then the master can only respond to a set command during the clock on-time period. Even if the master changes state during this on-time, the slave remains as it is. When the clock on-time ends, or goes from high to low, into the master, the clock into the slave goes from low to high, due to the inverter. At this time the master cannot accept data at either of its inputs because its clock line is low, or off. The slave, in effect, acts as a holding stage for data to be transferred to a next flip-flop (as in a shift register chain) and gives all cir-

78 Flip-Flops and Memories

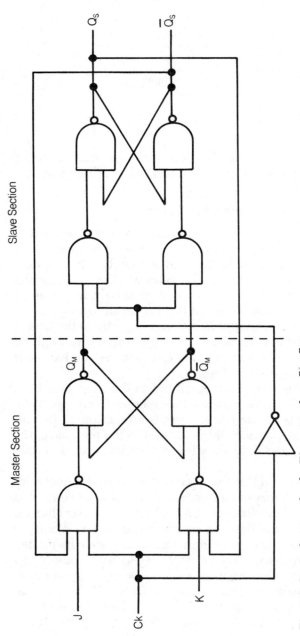

Fig. 5-7. Logic diagram for the JK master-slave flip-flop.

cuitry in the chain sufficient time to settle in. In a sense, the slave acts as a buffer for the data between the master and other flip-flops in more complex circuitry. The truth table for a JK master-slave flip-flop is shown in TABLE 5-6. Keep in mind that all of these types of flip-flops discussed are known as *digital memory devices*.

Table 5-6. JK Master-Slave Flip-Flop Truth Table

Ck	J	K	Q	$\overline{Q}$	Mode
0	X	X	NC	NC	Disable
0	0	0	NC	NC	Hold
0	1	0	1	0	Set
0	0	1	0	1	Reset
0	1	1	1/0	1/0	Toggle

SAMPLE-AND-HOLD

A sample-and-hold circuit is an analog memory. FIGURE 5-8 shows a basic circuit. A negative sampling pulse applied to the gate closes the circuit allowing the capacitor to charge to the instantaneous voltage of the input. In the absence of a pulse, the gate circuit opens with the capacitor retaining its charge. The output is thus a steady voltage level charging in steps between the sampling pulse intervals.

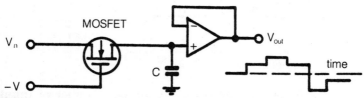

Fig. 5-8. Sample-and-hold, or analog memory.

To work effectively, the time of the sampling pulses must be short, the value of the capacitor low, and the output impedance of the op amp high in order not to discharge the capacitor between the sampling pulses. Also, the capacitor must be of a type which can hold its full charge between sampling pulses. Finally, a field effect transistor (MOSFET) is preferable to a bipolar transistor switch in most sample-and-hold applications, although the latter can be used. Today there are sample-and-hold ICs that contain the type of op amp applicable for this type of analog memory function.

READ-ONLY-MEMORY (ROM)

Read-only-memory (ROM) is a circuit which accepts a binary code (known as an *address*) at its input terminals and provides another binary code or word at its output terminals for each of the input combinations. Basically, therefore, it is a code-conversion system, although essentially it consists of a decoder applied to the input signals feeding an encoder providing the output signals. Since this encoder is essentially a memory matrix, the information it is provided with is stored and can be read out as often as required—hence the description read-only-memory (FIG. 5-9).

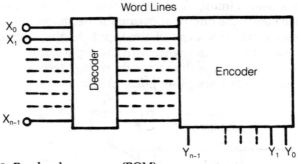

Fig. 5-9. Read-only memory (ROM).

Specifically, a ROM has a specified number of inputs (X_0, X_1, X_2, etc.) and a specified number of outputs (Y_0, Y_1, Y_2, etc.). These numbers are not necessarily the same. Thus, if there are X inputs and Y outputs, the capacity of that ROM is X words each of Y bits, or an X by Y bit memory. For example, if there are 32 inputs with 8 outputs this particular ROM has a capacity of 32 words each of 8 bits, or $32 \times 8 = 256$-bit memory. ROMs from 256-bit up to 1024-bit are typical figures for MSI using CMOS in conjunction with TTL logic. With LSI much larger memories can be achieved in a single package. Alternately, ROMs can be cascaded to provide larger memories. In the very near future you will probably see ROM memory devices with capacities in the megabytes (millions of bits of memory locations).

The way a ROM works is to decode the input into word lines (W_0, W_1, etc.), which are the minterm (see Appendix B) outputs of the decoder. These lines are then encoded again in the memory matrix where they are held. The working relationship can be established by a truth table or Boolean equations, or both, as a guide to implementation. Taking a four-input four-output ROM as a simple example, the truth table for conversion from binary code to a Gray code would look like TABLE 5-7. To accommodate different arithmetic codes some IC ROMs are designed to be programmable after manufacture (PROMs). Additionally, EPROMs and EEPROMs (erasable and electrically erasable PROMs respectively) provide a way to change the memory contents of ROMs from time to time as necessary.

RANDOM-ACCESS MEMORY (RAM)

A random-access memory (RAM) is a similar device to a ROM except that the stored words can be addressed and written directly as well as being read. RAM chips are sometimes also known as *read-and-write memory*. The decoder in this case employs latches (flip-flops) instead of diodes or transistors, which are bistable devices. This means that while a RAM provides stored memory, this is lost when the power supply is removed. For this reason, a RAM is described as a *volatile* device, with power dissipation necessary to maintain storage. In the case of certain types like a dynamic MOS RAM, a refreshing charge is necessary at regular intervals (every millisecond or so) to replace leakage of all capacitance on which the memory depends.

Table 5-7. Binary-To-Gray Code Conversion of a Four-Input/Four-Output ROM

Binary Inputs				Word Line	Gray Code Outputs			
X3	X2	X1	X0		Y3	Y2	Y1	Y0
0	0	0	0	W0	0	0	0	0
0	0	0	1	W1	0	0	0	1
0	0	1	0	W2	0	0	1	1
0	0	1	1	W3	0	0	1	0
0	1	0	0	W4	0	1	1	0
0	1	0	1	W5	0	1	1	1
0	1	1	0	W6	0	1	0	1
0	1	1	1	W7	0	1	0	0
1	0	0	0	W8	1	1	0	0
1	0	0	1	W9	1	1	0	1
1	0	1	0	W10	1	1	1	1
1	0	1	1	W11	1	1	1	0
1	1	0	0	W12	1	0	1	0
1	1	0	1	W13	1	0	1	1
1	1	1	0	W14	1	0	0	1
1	1	1	1	W15	1	0	0	0

and the corresponding Boolean equations would be:-

☐ Y0 = W1 + W2 + W5 + W6 + W9 + W10 + W13 + W14

☐ Y1 = W2 + W3 + W4 + W5 + W10 + W11 + W12 + W13

☐ Y2 = W4 + W5 + W6 + W7 + W8 + W9 + W10 + W11

☐ Y3 = W8 + W9 + W10 + W11 + W12 + W13 + W14 + W15

Dynamic MOS RAM

In a dynamic MOS RAM information can be stored on the parasitic gate-to-substrate capacitance, resulting in considerable circuit simplification where only three devices are needed to store four bits instead of the eight in a static MOS RAM. In this case, however, refreshing of all bits is required.

Typical IC RAM

A typical IC RAM is shown in FIG. 5-10. This figure shows the physical form (14 pin flat package) and block dia-

Random-Access Memory (RAM) 83

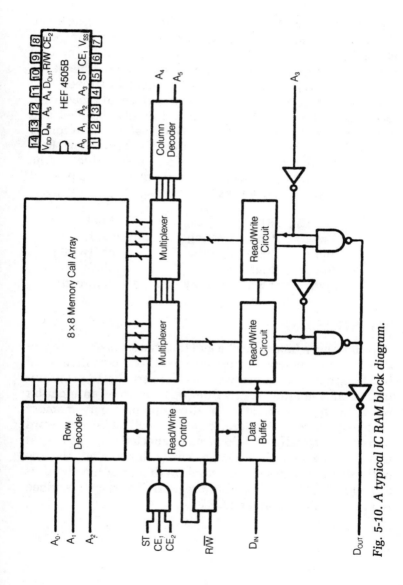

Fig. 5-10. A typical IC RAM block diagram.

gram for a 64-bit, 1-bit per word random access read/write memory. The memory is strobed for reading or writing only when the strobe input (ST), chip enable inputs (CE_1 and CE_2) are high simultaneously. The output data is available at the data output (D_{OUT}) only when the memory is strobed, the read/write input ($R/\overline{W}$) is high, and after the read access time has passed. Note that the output is initially disabled and always goes to the low state before data is valid. The output is disabled when the memory is not strobed or $R/\overline{W}$ is low. $R/\overline{W}$ may remain high during a read cycle or low during a write cycle. The output data has the same polarity as the input data. The function table is as follows:

ST, CE_1, CE_2	$R/\overline{W}$	D_{Out}	Mode
low	low	floating	disabled
high	low	floating	enabled(write)
low	high	floating	disabled
high	high	memory data	enabled(read)

REGISTERS

Flip-flops are a binary device and thus have a memory capacity of 1 bit of memory. It follows then that a combination of flip-flops can store as many bits as there are flip-flops, meaning, as an example, that 8 flip-flops can store an eight-bit word. Such a combination of flip-flops or binary memory devices is called a *register*. Normally, to allow the data word to be fed in serially, flip-flops are connected serially, output to input. The data is then progressively shifted along the line of flip-flops to complete the word. In this case, the circuit is referred to specifically as a *shift register*. These are described in more detail in chapter 11.

6

Number Systems

THERE are a number of different methods used to count in digital electronics. So far, you have been introduced to binary numbers and their use in representing decimal numbers. Binary numbers have a base of two, while decimal numbers have a base of 10. Knowing how to convert from one system to another can be very helpful in dealing with digital electronic circuits, but operations in digital circuitry can also be expressed using other number systems such as the *octal* number system and the *hexadecimal* number system. This last number system is especially important in understanding the operation of microprocessor circuitry, a field of study that is essential in understanding today's microprocessor based electronic equipment.

BINARY CODED DECIMALS

While digital electronic devices think, count, or react in terms of binary arithmetic (1 or 0, on or off), the human brain finds it much easier to think and communicate in decimal numbers. Some method of being able to render binary numbers in easily-readable decimal equivalents is therefore highly desirable, like an in-between system representing binary coded decimals.

This can be done quite simply. To represent the ten decimal numbers from 0 to 9, four binary digits or bits are required as shown below:

Decimal	Pure Binary			
	(2^3)	(2^2)	(2^1)	(2^0)
0	0	0	0	0
1	0	0	0	1
2	0	0	1	0
3	0	0	1	1
4	0	1	0	0
5	0	1	0	1
6	0	1	1	0
7	0	1	1	1
8	1	0	0	0
9	1	0	0	1

You can easily write decimal equivalents of 0 to 9 in separate groups of four bits, using as many groups as necessary to cover the number of digits in the decimal number. Taking the decimal number 7,893 as an example, each digit is treated separately as a number between 0 and 9:

decimal	7	8	9	3
binary coded decimal	0111	1000	1001	0011

This works equally as well the other way. To translate a binary coded decimal into its decimal equivalent each group is connected in turn:

binary coded decimal	0101	0011	1000	0111
decimal	5	3	8	7
or		5,387		

This particular system is known as an 8421 binary coded decimal, or 8421 BCD. The numbers here actually refer to the assigned values or weights given to the respective groups.

Using four groups, as in the example, the weights are:

$$2^3 = 8 \quad 2^2 = 4 \quad 2^1 = 2 \quad 2^0 = 1$$

A little further study shows that with this method of grouping, the four bits actually provide 16 possible combinations, only ten of which are used to cover the decimal numbers 0 to 9. In other words, six of the combinations are redundant, or unnecessary. This is shown in TABLE 6-1. As you can see, the decimal numbers 0 to 9 can be represented using 4 bits. Notice that 1001 is the largest 4-bit group in this 8421 code. This code does not use the numbers 1010, 1011, 1100, 1101, 1110, 1111. If any of these numbers appears in a digital machine using this code, an error has occurred.

Table 6-1. Decimal, 8421, and Binary Comparisons

Decimal	8421	Binary
0	0000	0000
1	0001	0001
2	0010	0010
3	0011	0011
4	0100	0100
5	0101	0101
6	0110	0110
7	0111	0111
8	1000	1000
9	1001	1001
10	0001 0000	1010
11	0001 0001	1011
12	0001 0010	1100
13	0001 0011	1101
14	0001 0100	1110
15	0001 0101	1111

As you can see, the 8421 code is the same as binary from 0 to 9. This is why it is called the 8421 code. However, with numbers greater than 9, the 8421 code is quite different from the pure binary number code. As an example, the binary number for 14 is 1110. But 0001 0100 is the 8421 code for the number 14. In the 8421 code, therefore, every number above 9 is very much different than numbers above 9 in the binary number code.

In a practical application, say using memory gates or flip-flops, to locate decimal numbers when fed by the 8421 BCD code (or vice versa), each group of numbers would need four flip-flops, and each set of four groups would have six unnecessary (redundant) code combinations. These could be eliminated by the use of a suitable alternative BCD.

TYPES OF CODES

There are many possible BCD code sequences, with the relative advantages of each depending on a variety of factors such as simplicity of circuit construction, operating speed, and ease of decoding for read-out purposes. Some are weighted codes while others are not.

Basic requirements of a weighted code are that the weights must be chosen so that their number is not greater than 15 and not less than 9. Additionally, one of the weights must be 1, and another either 1 or 2. For example, some possible combinations are 7421, 5421, 5211, 2421, and 8421 (already described). The respective group equivalents are shown in TABLE 6-2.

Table 6-2. Group Equivalents of Binary Numbers

Decimal	Pure Binary				Binary Coded Decimal			
	(2^3)	(2^2)	(2^1)	(2^0)	7421	5421	5211	2421
0	0	0	0	0	0000	0000	0000	0000
1	0	0	0	1	0001	0001	0001	0001
2	0	0	1	0	0010	0010	0100	0010
3	0	0	1	1	0011	0011	0110	0011
4	0	1	0	0	0100	0100	0111	0100
5	0	1	0	1	0101	1000	1000	0101
6	0	1	1	0	0110	1001	1001	0110
7	0	1	1	1	1000	1010	1011	0111
8	1	0	0	0	1001	1011	1110	1110
9	1	0	0	1	1010	1100	1111	1111

The 7421 BCD code has a particular advantage in practical applications in that it employs a minimum number of 1s. The figure 1 in a binary device represents an on state, normally drawing current. Thus, this code is attractive for providing minimum current consumption.

5421 BCD and 2421 BCD, or any other code where the sum of the weights is 9, yield the property that the 9's complement of the number (that is, $9-N$, where N is the number) can be obtained simply by inverting the binary equivalent. For example, in 5211 BCD, decimal 6 is given by 1001. Inverting this gives 0110 or the decimal number 3 ($9-6=3$). This again can be of particular advantage for certain types of circuits.

Three other codes are worth mentioning here. These are the *Excess Three Code*, the *Reflected* or *Gray Code*, and the *Johnson Code*. The Excess Three code is a self-complementing code obtained by adding 3 to each group of the binary code. It is very useful for performing decimal or binary coded decimal arithmetic. The Reflected or Gray code is also widely used, particularly in digital shift position encoders as it incurs only one digit change in passing from any one combination to the next. The Johnson code is quite different as this is an unweighted code, particularly adapted to counting because of the simplicity with which it can be decoded into decimal. TABLE 6-3 shows the equivalents in the three codes for decimals 0 to 9.

Table 6-3. Equivalent Numbers in Three Different Codes

Decimal	Pure Binary				Excess Three Code	Gray Code	Johnson Code
	(2^3)	(2^2)	(2^1)	(2^0)			
0	0	0	0	0	0011	0000	00000
1	0	0	0	1	0100	0001	00001
2	0	0	1	0	0101	0011	00011
3	0	0	1	1	0110	0010	00111
4	0	1	0	0	0111	0110	01111
5	0	1	0	1	1000	0111	11111
6	0	1	1	0	1001	0101	11110
7	0	1	1	1	1010	0100	11100
8	1	0	0	0	1011	1100	11000
9	1	0	0	1	1100	1101	10000

PARITY BITS

When the code used contains redundancies, the appearance of a redundancy number indicates an error. For example, the appearance of 1111 when using 8421 BCD indicates an error since no such number exists in the code. Errors produced by dropping or gaining a digit in the same code group, however, are not apparent as they still show valid combinations. The same is true of all codes used having no redundancies.

The simplest method of error detection is to add an extra bit, called a *parity bit*, in each group, giving this a value of 0 or 1 to make the total number of 1s in each group either odd or even. Should an error occur, this immediately shows up by the fact that the number of digits in the group will no longer be odd (or even).

The limitation of this is that only single errors show up. Two errors occurring in the same group return the sum of the digits to odd (or even) and show as correct. Three errors in the same group again indicate an error, but not whether a single or triple error occurred.

To check blocks of information the readout can be arranged in the form of a matrix. Parity checks are then made on the rows and columns, including the extra row formed by the column parity check (which also needs it own parity bit). An example of odd parity could be shown using the decimal number 8732. The normal 8421 BCD would be:

1000 0111 0011 0010

Adding an odd parity bit means that the total number of 1s in each four-bit group becomes an odd total. The decimal number 8732 with an odd parity bit added to the end of each four-bit group then becomes:

10000 01110 00111 00100

TABLE 6-4 shows the number 8732 in a manner in which the odd parity bit may be seen in a somewhat simpler manner. TABLE 6-5 is an example of the 8421 code for decimal

Table 6-4. An Example of Odd Parity

Decimal	BCD	Parity Bit	Total Bits
8	= ←1000	0	odd, OK
7	= ←0011	1	odd, OK
3	= ←0011	1	odd, OK
2	= ←0010	0	odd, OK

Table 6-5. Odd Parity for the 8421 Code

8421 Code	Added Bit
0000	1
0001	0
0010	0
0011	1
0100	0
0101	1
0110	1
0111	0
1000	0
1001	1

numbers 0 to 9 and the added bit that is necessary for odd parity.

In actual digital circuitry the probability of bit errors is actually very small. If an error does occur, it is most likely a one-bit error. However, because the possibility does exist, other methods of detecting multiple bit errors are used. One of these is the *Diamond code*. The Diamond code is designed to detect multiple errors using the property of all numbers which obey the formula 3n+2. The check is made by subtracting 2 from the combination and dividing the remainder by binary 3. If there is no remainder, the combination is valid.

OTHER NUMBER SYSTEMS

In most cases when you hear the word "number," you immediately think of decimal numbers. That is because you have learned to add, subtract, multiply, and divide in a number system that has a base of 10. Of course, you are now becoming more familiar with the binary number system but other methods used to represent numbers do exist. They are rooted in the binary number system but represent decimal numbers differently than does the 8421 code or the binary code. Two of these are discussed next.

Octal Numbers

The octal system is a numbering system with a base of 8. This means it has eight digits, 0 to 7, relative to the decimal system, although decimal 10 equals octal 8. The advantage of octal numbers is that they can be written as groups of three binary digits, called *binary triplets*. Thus, conversion from a binary number to an octal number is direct and straightforward as shown below:

Octal	Binary Triplet
0	000
1	001
2	010
3	011
4	100
5	101
6	110
7	111

To convert a binary number into its octal equivalent, the binary number is broken down into groups of three, or triplets. If necessary, zeros are added in front of the number to complete a set of triplets. The corresponding octal number then follows from the equivalent of the various triplets. An example is shown here:

```
Binary                         10110011
group in triplets              10 110 011
add zero to complete           010 110 011
corresponding octal numbers     2   6   3
Therefore the octal number equals 263₈
```

Notice that this is not the decimal number. To convert the octal number 263_8 to its decimal equivalent, remember that each digit in the octal system corresponds to a power of 8, just as in the binary system where each digit corresponds to a power of 2. In octal numbers, the weights of the digit positions are as follows:

$$8^3 \quad 8^2 \quad 8^1 \quad 8^0$$
(512 64   8   1   = decimal equivalents)

Therefore, to convert any octal number to a decimal number, multiply each octal digit by its weight and add the resulting products. In the case of the octal number 263_8, its decimal equivalent becomes:

$$2(8^2) + 6(8^1) + 3(8^0) = 128 + 48 + 24 = 200_{10}$$

Therefore, the decimal equivalent of the octal number 263 is 200.

Octal numbers can be used to check computer arithmetical solutions by comparing the answers obtained by the two numbering systems. A worked out example should make this clear.

Binary Sum	**Octal Sum**
110	6
+010	+2
1000	10
001 000	
1 0 octal equivalent	

The two octal numbers agree—the one derived directly by octal number working and the other extracted as the octal equivalent of the binary sum solution. Thus, the binary arithmetic is correct.

Hexadecimal Numbers

Hexadecimal numbers are numbers with a base of 16. After the number 9, letters A through F are used as shown in

TABLE 6-6. After the letter F, 2-digit combinations are used taking the second digit followed by the first digit, then the second digit followed by the second digit, and so on. This means that the next number following F is 10, then 11 to 19 followed by 1A to 1F, then 20 to 29, then 2A to 2F, and so on. The importance of hexadecimal numbers cannot be overstated. In the study and understanding of microcomputers, hex numbers are essential in programming, designing, and troubleshooting.

Table 6-6. Hexadecimal Conversion Table

Decimal	Binary	Hexadecimal
0	0000	0
1	0001	1
2	0010	2
3	0011	3
4	0100	4
5	0101	5
6	0110	6
7	0111	7
8	1000	8
9	1001	9
10	1010	A
11	1011	B
12	1100	C
13	1101	D
14	1110	E
15	1111	F

To convert a hex number to a binary number, simply convert each hex digit to its 4-bit equivalent. As an example, to convert 6BD to binary:

$$6 = 0110$$
$$B = 1011$$
$$D = 1101$$
$$\text{Therefore } 6BD_{16} = 0110\ 1011\ 1101$$

To convert a hex number to a decimal number you can either convert the hex number to a binary number first, then convert the binary number to the decimal number, or you can go directly from hex to decimal if you know the weights of the powers of 16. As an example, to convert C5F2 to decimal:

$$\begin{aligned} C5F2 &= C(16^3) + 5(16^2) + F(16^1) + 2(16^0) \\ &= 12(16^3) + 5(16^2) + 15(16^1) + 2(16^0) \\ &= 49{,}152 + 1{,}280 + 240 + 2 \\ &= 50{,}674 \end{aligned}$$

Most of today's scientific calculators are capable of converting from octal to binary, binary to decimal, and every combination in between. However, scientific calculators are not always readily available so it helps to know how to manually perform these conversion operations.

In the application of microcomputers, numbers made up of eight bits each are located in certain areas of memory known as addresses. If you want to retrieve a specific binary number from memory, you need to know its address. For a microcomputer that can store 65,536 eight-bit numbers, the address locations in binary would be 0000 0000 0000 0000 to 1111 1111 1111 1111. These would be address locations for the decimal numbers 0 through 65,535. But in hex, the addresses are 0000 through FFFF. As you can see, a great deal of time and energy can be saved using hex numbers.

HANDLING FRACTIONS

In the decimal system fractions are, of course, simply designated by a decimal point. Fractions are thus expressed in negative base values, such as 10^{-1}, 10^{-2}, 10^{-3}, etc. Exactly the same principle applies with any other numbering system although the resulting fractions will have quite different values. In the case of the binary system, for example, the negative base values are 2^{-1}, 2^{-2}, 2^{-3}, etc., the corresponding fractions being $1/2$, $1/4$, $1/8$, etc. Here are some typical comparisons:

Decimal

	10^0	10^{-1}	10^{-2}	10^{-3}
($1/8$)	0.	1	2	5
($1/4$)	0.	2	5	0
($1/2$)	0.	5	0	0
($3/4$)	0.	7	5	0
(1)	1.	0	0	0

Binary

2^0	2^{-1}	2^{-2}	2^{-3}
0.	0	0	0
0.	0	1	0
0.	1	0	0
0.	1	1	0
1.	0	0	0

Octal

	8^0	8^{-1}	8^{-2}
($1/8$)	0.	1	0
($1/4$)	0.	2	0
($1/2$)	0.	4	0
($3/4$)	0.	6	0
(1)	1.	0	0

One point which arises is that to express a fraction exactly, the denominator of the fraction must be exactly divisible by the base of the system. Thus the binary system can accommodate all fractions whose denominator is divisible by 2. It cannot, for example, accommodate $1/3$ as an exact value—nor can the decimal system ($1/3 = 0.3$ recurring).

7

Digital Clocks

THERE can be little doubt that timing circuits, called *clocks*, are an essential part of nearly all digital circuits. Clock circuits perform a number of functions. There is usually a master clock that is the source of pulse trains used to determine the speed at which a system operates, determines how long it takes to perform an operation such as addition, and ultimately controls all of the operations in the digital system. There are also subordinate clocks that use the master clock pulses as an input and provide output pulses that may be, depending upon their function, phase shifted or of a different frequency from the master clock. Clock circuits may also be referred to as *strobe circuits*. Basically, therefore, a clock as used in digital circuits is an oscillator which generates square waves or pulses (unlike a radio oscillator which generates sine waves). See FIGS. 7-1 through 7-4.

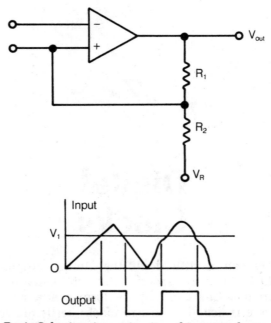

Fig. 7-1. Basic Schmitt trigger circuit and input and output characteristics. R1R2 is a voltage divider giving a feedback factor of R2/(R1 + R2).

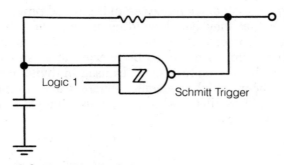

Fig. 7-2. Schmitt trigger pulse circuit.

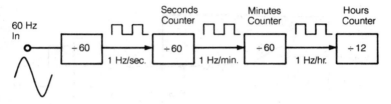

Fig. 7-3. Basic block diagram of frequency division for digital clock.

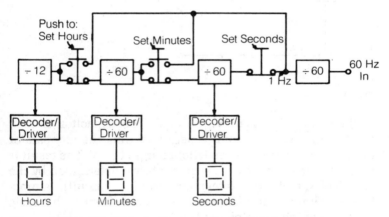

Fig. 7-4. Simplified block diagram of a digital clock.

OPERATIONAL AMPLIFIER CLOCKS

Op amps combined with an integrator can readily perform the clock function in some digital circuits as shown in FIG. 7-5. In this circuit the output is either $+V$ or $-V$. The op amp works as a comparator, comparing the input voltage V1 with a standard reference voltage V_R, V_1 being in the form of feedback from the voltage divider provided by R_2, R_3. If V_1 is more

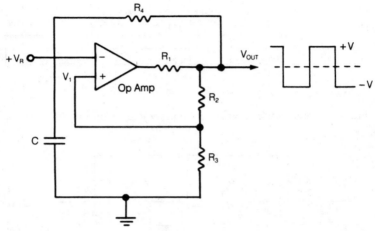

Fig. 7-5. Simple square wave generator.

positive than V_R, then $V_{out} = +V$. The capacitor C then charges to $+V$, when after a period of time the comparator output reverses and the capacitor charges to $-V$. The result is a square wave output with a time interval determined by the values of R_4 and C (the integrator part of the circuit). In practice, maximum pulse frequency obtainable from such a basic circuit is of the order of 10 kHz.

IC OSCILLATORS

A CMOS version of a ring oscillator is shown in FIG. 7-6. This clock circuit uses 4069 CMOS inverters connected in a loop or ring through R1 and R2. The frequency of this circuit is determined by the following equation:

$$f = \frac{1}{2C(0.405R_{eq} + 0.693R_1)}$$

where $R_{eq} = \dfrac{R1R2}{R1 + R2}$

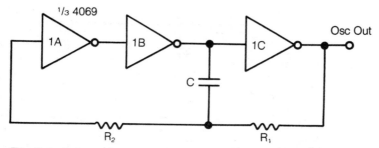

Fig. 7-6. A 4069 IC configured as a CMOS ring oscillator.

Some IC waveform generators provide square, triangular, and sine wave outputs simultaneously; one example is the L8038. It can also be phase locked to a reference. A working circuit for this IC is shown in FIG. 7-7. Square wave amplitude is of the order of 0.9 volts. The frequency is set by R and C and is calculated by:

$$\text{frequency} = \frac{0.15}{R \times C}$$

An alternative IC which provides both square and triangular waveform outputs is shown in FIG. 7-8. Here the frequency is given by:

$$\text{frequency} = \frac{2(+V - V_{cc})}{R_1 \times C_1 \times V_{cc}}$$

MONOSTABLE MULTIVIBRATORS

Multivibrators are analog rather than digital devices, but they are readily capable of working as pulse generators. A monostable multivibrator has one stable state and one quasi-stable state. Starting in its stable state, a triggering signal transforms it into its quasi-stable state when, after a certain period of time, the circuit returns to its stable state. Thus the output is in the form of a pulse width equal to the circuit

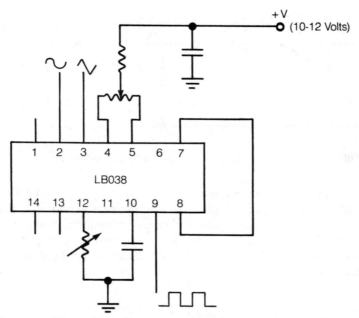

Fig. 7-7. IC waveform generator providing sine, triangular, or square wave outputs.

delay time. A further triggering signal is necessary to generate another pulse, and so on. Because of this, it is known as a one-shot multivibrator.

A practical circuit is shown in FIG. 7-9. This is a 555 timer connected as a monostable multivibrator. For each negative clock edge input, one positive output pulse is produced. R and C in this circuit determine the width of the output pulse which may be calculated using the following equation:

$$TH = 1.1RC$$

Monostable multivibrators (one-shots) are also available in single function ICs. These are the 74121, 74122, and 74123. A circuit using two op amps and capable of generating both positive and negative pulses is shown in FIG. 7-10.

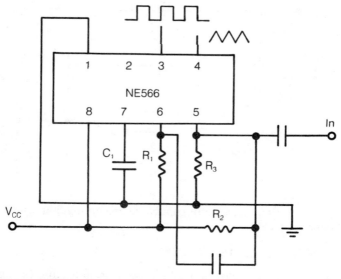

Fig. 7-8. IC waveform generator providing square or triangular waveform outputs.

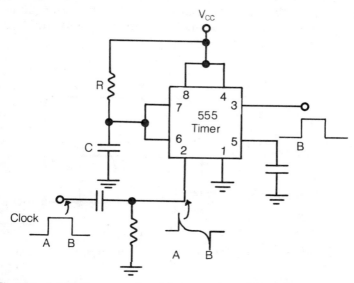

Fig. 7-9. A 555 timer configured as a monostable multivibrator.

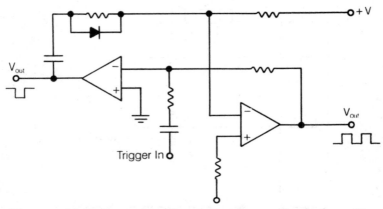

Fig. 7-10. Multivibrator circuit capable of generating both positive and negative pulses.

BISTABLE MULTIVIBRATORS

A bistable multivibrator is stable in both its states and is generally known as a flip-flop. It is a true digital rather than analog device, which in a sequential circuit is set and reset by clock pulses. An example of a practical bistable multivibrator circuit is shown in FIG. 7-11 based on an op amp and five resistors.

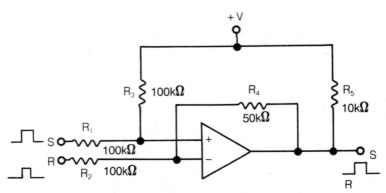

Fig. 7-11. Bistable (digital) multivibrator using an LM193 op amp.

CRYSTAL CONTROLLED OSCILLATORS

An op amp can be used to construct a crystal controlled oscillator as shown in FIG. 7-12. Here, the crystal is used in parallel with a capacitive voltage divider. At resonance (oscillator fre-

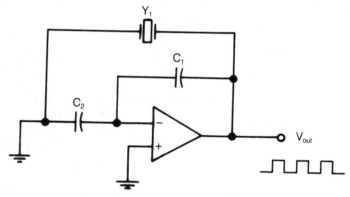

Fig. 7-12. A crystal-controlled oscillator using an op amp as the active device.

quency), the impedance of the feedback tank circuit is maximum with the excitation voltage of the crystal being determined by the ratio of C_1 to C_2. If the output frequency of the amplifier drifts, the impedance of the crystal decreases, shunting the undesired frequencies to ground.

SWEEP GENERATORS

A linear amplifier (op amp) used in conjunction with a resistor and a capacitor can be made to work as a triangle wave generator by the use of integration. This is shown in FIG. 7-13. If the input is a constant voltage, the output is in the form of a linear ramp or sweep waveform. Although this is a linear device it can be used in hybrid circuits, so it is worthy of brief description. Figures 7-14 and 7-15 show two other examples of sweep generators.

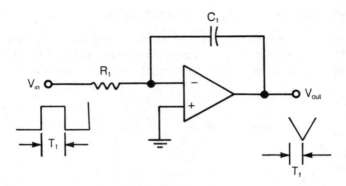

Fig. 7-13. A basic linear ramp or sweep waveform generator.

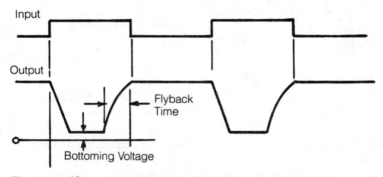

Fig. 7-14. Characteristic output of a Miller sweep generator.

Sweep Generators 107

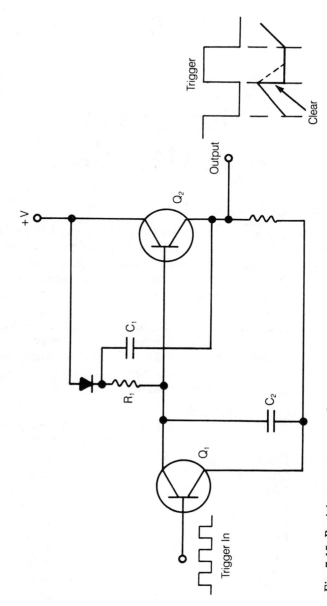

Fig. 7-15. Bootstrap sweep generator.

8

Encoders and Decoders

ENCODERS

A binary encoder consists of a suitable number of inputs, each of which represents a line in the binary code involved. It then provides direct access to any one line whereby an input signal applied to that line gives a 1 output, or generates a bit. A binary encoder converts a decimal number to a binary number.

Suppose the binary code has to cover a count of 10 decimal, meaning it is required to have 10 bits. This can only be satisfied with a minimum of $2^4 = 16$ bits ($2^3 = 8$ is not enough), of which $16 - 10 = 6$ is redundant since only 10 lines are required. These are shown below:

	Decimal	Output Code			
		Bit 3 (2^3)	Bit 2 (2^2)	Bit 1 (2^1)	Bit 0 (2^0)
Line 0	0	Y3	Y2	Y1	Y0
Line 1	1	0	0	0	1
Line 2	2	0	0	1	0
Line 3	3	0	0	1	1
Line 4	4	0	1	0	0
Line 5	5	0	1	0	1
Line 6	6	0	1	1	0
Line 7	7	0	1	1	1
Line 8	8	1	0	0	0
Line 9	9	1	0	0	1

Encoders and Decoders   109

This can be encoded in the form of a diode matrix, which for simplicity is shown as a wired circuit with keys for each input, with outputs Y0, Y1, Y2, Y3 indicating the state of the matrix via lamps as shown in FIG. 8-1. To complete any line circuit to its corresponding lamp current must flow through a diode to provide OR logic. In the absence of a diode in the circuit cleared by any key, the corresponding

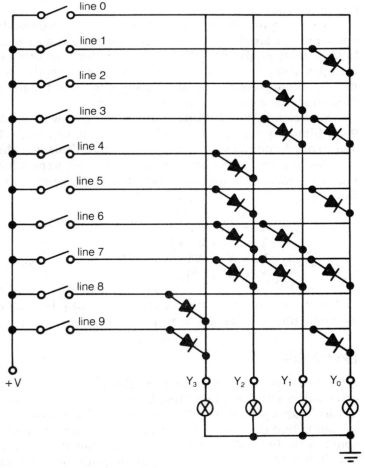

Fig. 8-1. Diode matrix encoder.

vertical or output line is at signal 0. As you can see this only occurs at key 1 position, representing decimal 0. Equally, each diode can be replaced by a transistor working as a diode (base and emitter connections), with the advantage that only one multiple-emitter transistor is required instead of fifteen diodes. In practice, several transistors may be needed for coverage, depending on the number of bits in the output code. The number of emitters required is equal to the number of bits in the code.

Assuming that the keys are not operated simultaneously, operation of a single key encodes the decimal number position in binary equivalent (all other lines at this time being in the open state). For example, closing key 8 (to encode decimal 7), output lines Y2, Y1, and Y0 are actuated (through the diodes) giving a complete output signal 0111.

Specifically, Y0 = 1 if line 1, line 3, line 5, line 7, or line 9 = 1. Similarly Y1 = 1 if line 2, line 3, line 6, or line 7 = 1, and so on. A complete truth table is shown in TABLE 8-1. This can also be expressed in Boolean algebra as such (remembering that + means OR logic and that the letter W represents a line):

$$Y0 = W1 + W3 + W5 + W7 + W9$$
$$Y1 = W2 + W3 + W6 + W7$$
$$Y2 = W4 + W5 + W6 + W7$$
$$Y3 = W8 + W9$$

Such an encoding matrix, therefore, can be implemented with OR gates and diodes.

DECODERS

A decoder is a system whereby digital information is extracted in a different form; that is, a binary code to be read in decimal equivalent (BCD-to-decimal decoder). Again assuming that the binary unit is a four-bit device (as with a count of decimal 10) a basic decoder to cover this requires four inputs (A, B, C, and D) and ten output lines (covering decimal 0 to 9).

Table 8-1. Encoder Truth Table

Inputs (Lines)										Outputs			
9	8	7	6	5	4	3	2	1	0	Y3	Y2	Y1	Y0
0	0	0	0	0	0	0	0	0	1	0	0	0	0
0	0	0	0	0	0	0	0	1	0	0	0	0	1
0	0	0	0	0	0	0	1	0	0	0	0	1	0
0	0	0	0	0	0	1	0	0	0	0	0	1	1
0	0	0	0	0	1	0	0	0	0	0	1	0	0
0	0	0	0	1	0	0	0	0	0	0	1	0	1
0	0	0	1	0	0	0	0	0	0	0	1	1	0
0	0	1	0	0	0	0	0	0	0	0	1	1	1
0	1	0	0	0	0	0	0	0	0	1	0	0	0
1	0	0	0	0	0	0	0	0	0	1	0	0	1

To accommodate all possible input states, eight inputs are required; A, $\overline{A}$, B, $\overline{B}$, C, $\overline{C}$, D, $\overline{D}$. To cover ten output lines, ten four-input NAND gates are needed. The basic circuit is then as shown in FIG. 8-2. (In practice the complementary inputs $\overline{A}$, $\overline{B}$, $\overline{C}$, and $\overline{D}$ may be obtained using inverters.) This circuit then works in the opposite manner to a decoder; the outputs and inputs are transposed. The truth table for the decoder of FIG. 8-2 is shown in TABLE 8-2.

For example, a binary input $\overline{A}B\overline{C}D$ or 0101 gives an immediate output on line 5 (decimal 5). These requirements can also be implemented by a diode matrix working with AND logic.

MULTIPLEXERS

A multiplexer lets you select 1 out of any number of input sources, directing this data to a single information channel. It is normally specified by an N-to-1 multiplexer, N being the number of inputs it is designed to select from. A typical basic circuit for a 4-to-1 multiplexer is shown in FIG. 8-3 using AND gates and AND-OR logic.

DEMULTIPLEXERS

A demultiplexer performs the inverse function of a multiplexer. It provides a binary signal on any one of N lines to

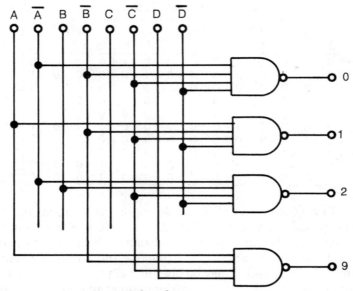

Fig. 8-2. BCD-to-decimal decoder.

Table 8-2. Decoder Truth Table

Inputs				Output Lines									
A	B	C	D	9	8	7	6	5	4	3	2	1	0
0	0	0	0	0	0	0	0	0	0	0	0	0	1
0	0	0	1	0	0	0	0	0	0	0	0	1	0
0	0	1	0	0	0	0	0	0	0	0	1	0	0
0	0	1	1	0	0	0	0	0	0	1	0	0	0
0	1	0	0	0	0	0	0	0	1	0	0	0	0
0	1	0	1	0	0	0	0	1	0	0	0	0	0
0	1	1	0	0	0	0	1	0	0	0	0	0	0
0	1	1	1	0	0	1	0	0	0	0	0	0	0
1	0	0	0	0	1	0	0	0	0	0	0	0	0
1	0	0	1	1	0	0	0	0	0	0	0	0	0

which it is addressed. It can be derived directly from a decoder by the addition of a signal (S) line as shown in FIG. 8-4. When a data signal is applied at S, the output appears only on the addressed line as the complement of this signal.

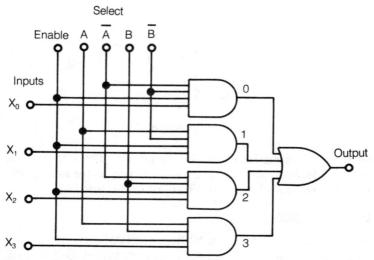

Fig. 8-3. Basic 4-to-1 multiplexer circuit.

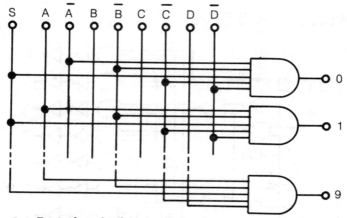

Fig. 8-4. Basic demultiplexer with S input.

In practice, this working is normally combined with an inhibit or enable input (also called a strobe input) feeding the S terminal as shown in FIG. 8-5. In this case if the enable input is 1, the data is inhibited from appearing on any line. If both data and enable inputs are 0, the data appears directly

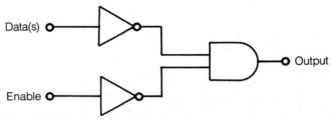

Fig. 8-5. Inhibit or enable input function.

on the addressed line without inversion. The capacity of demultiplexers is specified in the same way as for multiplexers such as 2-to-4 line, 3-to-8 line, or 4-to-6 line.

IC DECODERS

A practical example of an IC decoder is shown in FIG. 8-6. This has four inputs to accept a four-bit binary coded decimal (8421 BCD code) and 10 outputs (O_0, O_1, etc.). The truth

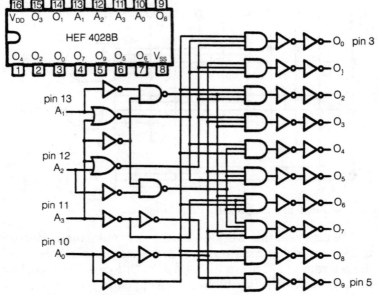

Fig. 8-6. A typical CMOS IC decoder.

table, written in terms of H=high state or a more positive voltage signal and L=low state or a less positive voltage, is shown in TABLE 8-3.

Basically, an 8421 BCD code applied to the inputs causes the selected output to be H, and the other source L. This device can also be used as a 1-of-8 decoder with enable. In this case three-bit octal inputs are applied to A_0, A_1, and A_2, selecting an output from O_0 to O_7. Input A_3 then becomes an active LOW enable forcing the selected output to L when A_3 is H.

Table 8-3. IC Decoder Truth Table

Inputs				Outputs									
A_3	A_2	A_1	A_0	O_0	O_1	O_2	O_3	O_4	O_5	O_6	O_7	O_8	O_9
L	L	L	L	H	L	L	L	L	L	L	L	L	L
L	L	L	H	L	H	L	L	L	L	L	L	L	L
L	L	H	L	L	L	H	L	L	L	L	L	L	L
L	L	H	H	L	L	L	H	L	L	L	L	L	L
L	H	L	L	L	L	L	L	H	L	L	L	L	L
L	H	L	H	L	L	L	L	L	H	L	L	L	L
L	H	H	L	L	L	L	L	L	L	H	L	L	L
L	H	H	H	L	L	L	L	L	L	L	H	L	L
H	L	L	L	L	L	L	L	L	L	L	L	H	L
H	L	L	H	L	L	L	L	L	L	L	L	L	H
H	L	H	L	L	L	L	L	L	L	L	L	H	L
H	L	H	H	L	L	L	L	L	L	L	L	L	H
H	H	L	L	L	L	L	L	L	L	L	L	H	L
H	H	L	H	L	L	L	L	L	L	L	L	L	H
H	H	H	L	L	L	L	L	L	L	L	L	H	L
H	H	H	H	L	L	L	L	L	L	L	L	L	H

Note: X = indifferent or "doesn't care" state

1-OF-16 DECODER/DEMULTIPLEXER

The HEF4515B 1-of-16 decoder/demultiplexer is an excellent example of how much logic can be contained in a small IC package. This has four binary weighted address inputs (A_0, A_1, A_2, and A_3) and 16 outputs, a latch enable input (EL), and an active LOW enable input $\overline{E}$. When EL is HIGH the selected output is determined by the data on A_0 to A_3. When EL goes

116 Encoders and Decoders

LOW the last data present are stored (latched) and the outputs remain stable. When $\overline{E}$ goes LOW, the selected output is determined by the contents if the latch is LOW, and when $\overline{E}$ goes HIGH, all outputs are HIGH. The enable input $\overline{E}$ does not affect the state of the latch. The logic diagram for this device is shown in FIG. 8-7. The corresponding truth table (EL HIGH) is shown in TABLE 8-4.

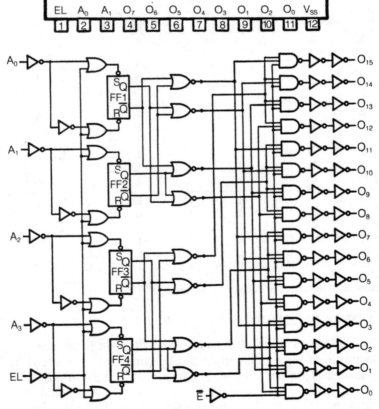

Fig. 8-7. 1-of-16 decoder/demultiplexer.

1-of-16 Decoder/Demultiplexer

Table 8-4. 1-of-16 Decoder/Demultiplexer Truth Table

Inputs					Outputs															
Ē	A₀	A₁	A₂	A₃	O₀	O₁	O₂	O₃	O₄	O₅	O₆	O₇	O₈	O₉	O₁₀	O₁₁	O₁₂	O₁₃	O₁₄	O₁₅
H	X	X	X	X	H	H	H	H	H	H	H	H	H	H	H	H	H	H	H	H
L	L	L	L	L	L	H	H	H	H	H	H	H	H	H	H	H	H	H	H	H
L	H	L	L	L	H	L	H	H	H	H	H	H	H	H	H	H	H	H	H	H
L	L	H	L	L	H	H	L	H	H	H	H	H	H	H	H	H	H	H	H	H
L	H	H	L	L	H	H	H	L	H	H	H	H	H	H	H	H	H	H	H	H
L	L	L	H	L	H	H	H	H	L	H	H	H	H	H	H	H	H	H	H	H
L	H	L	H	L	H	H	H	H	H	L	H	H	H	H	H	H	H	H	H	H
L	L	H	H	L	H	H	H	H	H	H	L	H	H	H	H	H	H	H	H	H
L	H	H	H	L	H	H	H	H	H	H	H	L	H	H	H	H	H	H	H	H
L	L	L	L	H	H	H	H	H	H	H	H	H	L	H	H	H	H	H	H	H
L	H	L	L	H	H	H	H	H	H	H	H	H	H	L	H	H	H	H	H	H
L	L	H	L	H	H	H	H	H	H	H	H	H	H	H	L	H	H	H	H	H
L	H	H	L	H	H	H	H	H	H	H	H	H	H	H	H	L	H	H	H	H
L	L	L	H	H	H	H	H	H	H	H	H	H	H	H	H	H	L	H	H	H
L	H	L	H	H	H	H	H	H	H	H	H	H	H	H	H	H	H	L	H	H
L	L	H	H	H	H	H	H	H	H	H	H	H	H	H	H	H	H	H	L	H
L	H	H	H	H	H	H	H	H	H	H	H	H	H	H	H	H	H	H	H	L

Note: X = indifferent or "doesn't care" state

LED READOUT

LED displays are commonly used to provide visual readout of digital information in decimal numbers. Each number requires a seven-segment light-emitting diode (LED) to cover numbers from 0 to 9 as shown in FIG. 8-8. Thus, it is necessary to convert a digital input of binary coded decimal form into a 7-bit (7-segment) display code. This is shown in the truth table of TABLE 8-5. As an example, to display decimal 6, the binary code 0110 has to be converted into output code 1111100, powering segments Y6, Y5, Y4, Y3, and Y2, with segments Y1 and Y0 off.

Fig. 8-8. Seven-segment LED display.

Table 8-5. Seven-Segment Display Code Truth Table

Decimal Number	D3	D2	D1	D0	Word Line	7-Bit Output Code						
						Y6 (g)	Y5 (f)	Y4 (e)	Y3 (d)	Y2 (c)	Y1 (b)	Y0 (a)
0	0	0	0	0	W0	0	1	1	1	1	1	1
1	0	0	0	1	W1	0	0	0	0	1	1	0
2	0	0	1	0	W2	1	0	1	1	0	1	1
3	0	0	1	1	W3	1	0	1	1	1	1	1
4	0	1	0	0	W4	1	1	0	0	1	1	0
5	0	1	0	1	W5	1	1	0	1	1	0	1
6	0	1	1	0	W6	1	1	1	1	1	0	0
7	0	1	1	1	W7	0	0	0	0	1	1	1
8	1	0	0	0	W8	1	1	1	1	1	1	1
9	1	0	0	1	W9	1	1	0	0	1	1	1

One method of accomplishing this is to use ROM as a code converter, specifically providing the required output code. LEDs draw only small amounts of currents and so can be powered directly from the IC output. Since four inputs are required, the ROM would actually provide 16 possible input combinations, six of which are unused. It is possible, however, to design a minimized converter circuit with no redundancies and such devices are available in IC form specifically for such applications. They are called N-segment/decoder/drivers, where N is the number of LED segments covered.

LED displays are themselves produced in packaged form, normally with 14 pins to fit standard sockets. Some examples of internal wiring are shown in FIG. 8-9. These are 7-segment digit displays plus a decimal point.

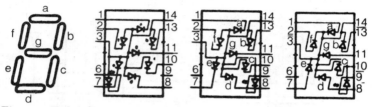

Fig. 8-9. Typical internal wiring of seven-segment LED displays.

A series of such displays, used to read out more than one digit, normally has a common connection (common cathode or common anode), when the basic circuitry involved is as shown in FIG. 8-10. Arrays of this type, of course, are not restricted to numerals. They can present readout in letters (such as A, B, C, D, E, F, G, H, etc.), mixed figures and numbers, or other symbols, although not all available combinations with a 7-segment LED cover the full alphabet. A similar device can also be used to power liquid crystal displays.

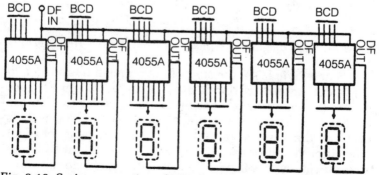

Fig. 8-10. Series connection of LED displays with common cathode (or common anode).

DISPLAY DRIVERS

An example of an LED display driver logic diagram is shown in FIG. 8-11 which is packaged in the form of a 16-lead DIP IC. This has four address inputs (coded D_A to D_D) and seven outputs (coded O_a to O_g). In addition, this is an active LOW latch enable input ($\overline{EL}$), an active LOW ripple blanking input ($\overline{BI}$), and an active LOW lamp test input ($\overline{LT}$).

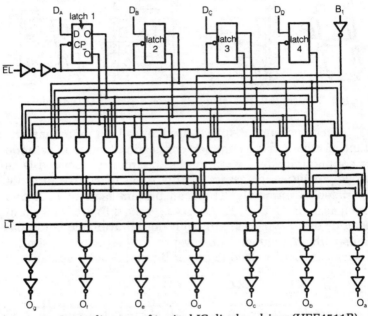

Fig. 8-11. Logic diagram of typical IC display driver (HEF4511B).

When $\overline{EL}$ is LOW, the state of the segment outputs (O_a to O_g) is determined by the data on D_A to D_D. When $\overline{EL}$ goes HIGH, the last data present on D_A to D_D are stored in the latches and the segment outputs remain stable. When $\overline{LT}$ is LOW, all the segment outputs are HIGH independent of all other input conditions. With $\overline{LT}$ HIGH, a LOW on $\overline{BI}$ forces all segment outputs LOW. The inputs $\overline{LT}$ and $\overline{BI}$ do not affect the latch circuit.

Encoders and Decoders   121

Table 8-6. Full Function Table of HEF4511B Display Driver

EL	BI	LT	D_D	D_C	D_B	D_A	O_a	O_b	O_c	O_d	O_e	O_f	O_g	Display
X	X	L	X	X	X	X	H	H	H	H	H	H	H	8
X	L	H	X	X	X	X	L	L	L	L	L	L	L	Blank
L	H	H	L	L	L	L	H	H	H	H	H	H	L	0
L	H	H	L	L	L	H	L	H	H	L	L	L	L	1
L	H	H	L	L	H	L	H	H	L	H	H	L	H	2
L	H	H	L	L	H	H	H	H	H	H	L	L	H	3
L	H	H	L	H	L	L	L	H	H	L	L	H	H	4
L	H	H	L	H	L	H	H	L	H	H	L	H	H	5
L	H	H	L	H	H	L	L	L	H	H	H	H	H	6
L	H	H	L	H	H	H	H	H	H	L	L	L	L	7
L	H	H	H	L	L	L	H	H	H	H	H	H	H	8
L	H	H	H	L	L	H	H	H	H	L	L	H	H	9
L	H	H	H	L	H	L	L	L	L	L	L	L	L	Blank
L	H	H	H	L	H	H	L	L	L	L	L	L	L	BlanK
L	H	H	H	H	L	L	L	L	L	L	L	L	L	Blank
L	H	H	H	H	L	H	L	L	L	L	L	L	L	Blank
L	H	H	H	H	H	L	L	L	L	L	L	L	L	Blank
L	H	H	H	H	H	H	L	L	L	L	L	L	L	Blank

Note: X = indifferent or "doesn't care" state

In this description, HIGH corresponds to a signal 1 and LOW to a signal 0. The full function table (truth table plus the other input functions) is shown in TABLE 8-6. Input conditions marked X indicate that the state is immaterial or "don't care."

9

Digital Adders

BINARY ADDERS

Binary adders perform the mathematical operation of addition using *bits* (binary digits). They can also be used to perform subtraction (negative addition), multiplication (repeated addition), and division (repeated subtraction) by suitable programming. In other words, all the common arithmetical functions can be performed by binary adders, which in turn are a basic application of logic gates.

HALF-ADDERS

A two-input device known as a *half-adder* (HA), has to cope with $2^2 = 4$ possible combinations of input signals and provide a realistic output. This means coverage of all input conditions in a meaningful way. To do this it must have two outputs, one to provide a readout for the addition within the capability of a two-digit count (0 and 1), and the other to accommodate overflow or carry to another counting stage. The half-adder then, can sum two binary digits and pass on the result, with a remainder. It cannot, however, accommodate a third digit, carried over from a previous sum.

Calling the inputs A and B and the outputs R (signal readout or display) and C (carry), the truth table is as shown in TABLE 9-1. As you can see, while three of the combinations

124 Digital Adders

Inputs		Outputs	
A	B	R	C
0	0	0	0
0	1	1	0
1	0	1	0
1	1	0	1

Table 9-1. Half-Adder Truth Table

producing the sum can be represented by a single digit readout either as 1 or 0, the condition 11 cannot. It represents an overflow condition; hence, the readout must revert to 0 with a carry of 1.

In terms of logic gates, the first three combinations can be covered by an exclusive OR gate. To accommodate the carry, an AND gate must be added as shown in FIG. 9-1. The output of the exclusive-OR gate is the sum, and the output of the AND gate is the carry.

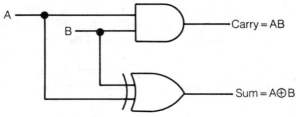

Fig. 9-1. Half-adder (HA) circuit.

FULL-ADDERS

To extend addition to accommodate more digits (starting by accommodating the carry from a ha..-adder), half-adders can be cascaded to make a full-adder. This is shown in FIG. 9-2 with provision to accept two inputs A and B directly into this stage and a carry input from the initial stage (which need only be a half-adder). To provide the facility to carry C_1 or C_2 forward to the next stage (there cannot be a carry output at both C_1 and C_2 simultaneously) the carry output must be taken through an OR gate.

In practice, full-adders are not necessarily constructed from two half-adders. The number of components required to

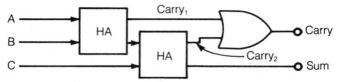

Fig. 9-2. *Full-adder circuit from half-adders.*

produce the required function can be reduced since only seven signal combinations are required, as defined by truth TABLE 9-2 for the full-adder. The Boolean equations corresponding to the truth table are:

$$Sn = \overline{A}nBnC_{n-1} + \overline{A}nBn\overline{C}_{n-1} + An\overline{B}n\overline{C}_{n-1} + AnBnC_{n-1}$$
$$Cn = \overline{A}nBnC_{n-1} + An\overline{B}nC_{n-1} + AnBn\overline{C}_{n-1} + AnBnC_{n-1}$$

Table 9-2. *Full-Adder Truth Table*

Inputs			Outputs	
A	B	C_{in}	C_{out}	S
0	0	0	0	0
0	0	1	1	0
0	1	0	1	0
0	1	1	0	1
1	0	0	1	0
1	0	1	0	0
1	1	0	0	1
1	1	1	1	1

These equations represent what is referred to as a *sum of products*; hence, each term in the equation is called a *minterm*. Considered as minterms, the equations can readily be simplified to:

$$Sn = An\overline{C}n + Bn\overline{C}n + AnBnCn - 1 + Cn - 1\overline{C}n$$
$$Cn = AnBn + BnC_{n-1} + AnC_{n-1}$$

An example of implementing these simplified equations in hardware form using AND and OR gates is shown in FIG. 9-3.

FIGURE 9-4 then shows a four-bit full-adder capable of reading (or displaying) up to a maximum count of $2^3 = 8$ in

126 Digital Adders

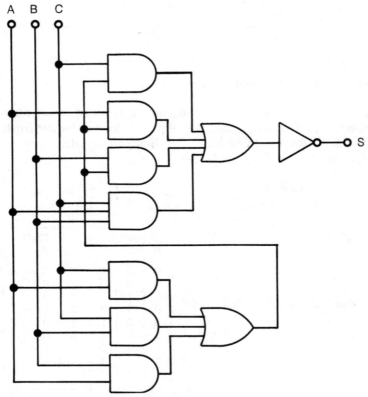

Fig. 9-3. Gate circuit for full-adder based on minterms.

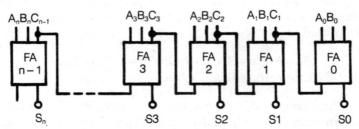

Fig. 9-4. Cascaded full-adders.

binary numbers. Note that the working is from right to left, but appropriate inputs can be made to any stage directly, keeping in mind that they do all have to be fed through the first stage as a series of 1 signals. The circuit is an adder, not a counter. Also the first (right hand) stage does not have to be a full-adder, only a half-adder (although in a practical IC it is usually a full-adder with the third input not used).

Obviously the coverage can be extended by adding further full-adders to the left. Commercial IC binary adders are generally available with one-bit, two-bit, and four-bit coverage (sometimes more), depending on the number of pins available. A four-bit adder requires 16 pins; 8 for inputs, 4 for sum outputs, 1 for carry output, 1 for carry input (to allow this IC to be cascaded with other full adders), 1 for power input, and 1 for ground. Carry connections are completed internally.

BINARY SUBTRACTORS

The basic rule of binary subtraction is to add the binary complement of the number to be subtracted. In practice this involves an extra bit being introduced which may be subject to what is referred to as *end carry round*. For example, to subtract a four-bit number B from another four-bit number A, the solution is to add A, B, and 1. The basic circuitry, as applied to a four-bit adder to turn it into a subtractor, is as shown in FIG. 9-5. The basic functions involved for a four-bit subtractor are:

$$B \text{ plus } \overline{B} = 1111$$
$$B \text{ plus } \overline{B} \text{ plus } 1 = 10000$$

Hence, $B = 10000$ minus $\overline{B}$ minus 1 when A minus B = A plus $\overline{B}$ minus 1000. The 1 is the output carry C_{out} fed back to the carry input C_{in}. This works as long as A is greater than B, yielding a positive difference. If B is greater than A, yielding a negative difference, there is no carry round and a

128 Digital Adders

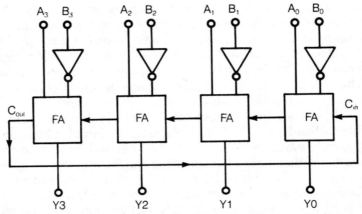

Fig. 9-5. Basic circuitry for a four-bit (parallel) binary subtractor with end-around carry.

slightly different system must be used. In practice, an IC adder/subtractor incorporates a true/complement unit to handle both positive and negative differences, as shown in FIG. 9-6. In the case of a negative difference, the correct solution is then obtained by complementing the sum digits S_0, S_1, S_2, and S_3. In the case of a positive difference, there is a carry and the solution is given directly by the S_0, S_1, S_2, and S_3 bits.

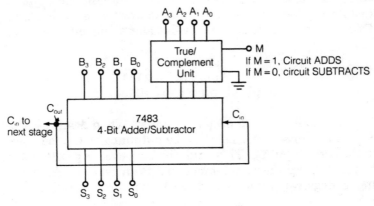

Fig. 9-6. IC adder/subtractor with true/complement unit.

SERIAL ADDER/SUBTRACTOR

In the case of a serial adder, the inputs are synchronous pulse trains applied to the individual lines. The output is then either the combined waveform of the inputs (addition) or the difference. This can be performed by a single full-adder, with carry facility for subtraction and a time delay in the carry line to inject the carry pulse (when present) into the digit pulses at the correct time interval. (See FIG. 9-7.)

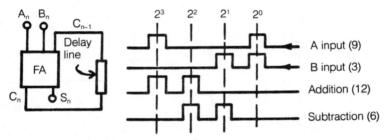

Fig. 9-7. Serial adder/subtractor.

The chief advantage of a serial adder/subtractor is that only a minimum of components are required, only one FA and a time delay. It is slower than the previous type described, which uses parallel working, but at the expense of requiring one full-adder for each bit. Various other types of circuits may be employed for adders, particularly one for binary coded decimals called an 8421 adder. A circuit used for BCD operation is shown in FIG. 9-8.

This circuit adds 8421 digits using binary addition. When the sum exceeds 9, a correction of 0110 is added. You may want to try adding two numbers such as 7 (0111) and 6 (0110) to see for yourself how this circuit works. There is also another adder/subtractor called a *2's complement adder/subtractor* shown in FIG. 9-9. Recall that the adder/subtractor that was first discussed is a 1's complement type. This other adder/subtractor is slightly different. When SUB (subtract) is low or 0, the B bits pass through the exclusive-OR (controlled inverters) to the full-adders. Hence, the full-adders produce the sum of A and B. However, when SUB is high or 1, the B

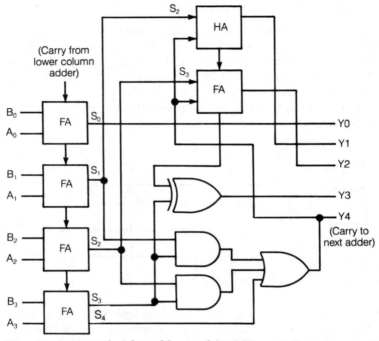

Fig. 9-8. An 8421 four-bit adder used for BCD operation.

bits are inverted before reaching the full-adders. This 1 is also initially added to the first full-adder forming the 2's complement of B. The output, therefore, of the full-adders is the difference of A and B.

HALF AND FULL-SUBTRACTORS

Half-subtractors and full-subtractors can also be used directly without taking the complements of binary numbers. Recall that binary numbers can be subtracted using the following rules:

$0-0=0$ with a borrow of 0
$0-1=1$ with a borrow of 1

Half and Full-Subtractors

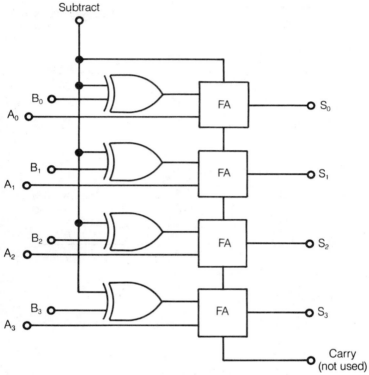

Fig. 9-9. A 2's complement adder/subtractor.

$$1-0=1 \quad \text{with a borrow of 0}$$
$$1-1=0 \quad \text{with a borrow of 0}$$

As you can see, if a circuit could be designed to produce both the borrow and difference outputs, complements would not have to be used. This is accomplished using the circuit of FIG. 9-10, known as a *half-subtractor*. A *full-subtractor* is shown in FIG. 9-11 and a *parallel 4-bit binary subtractor* is shown in FIG. 9-12.

Arithmetic in computers relies on the fundamental logic gates that you have already been studying.

132 Digital Adders

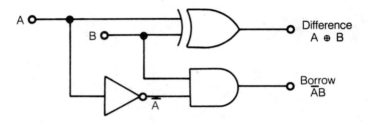

Fig. 9-10. A half-subtractor.

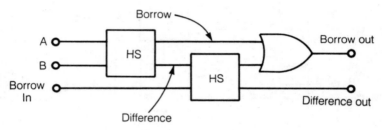

Fig. 9-11. A full-subtractor.

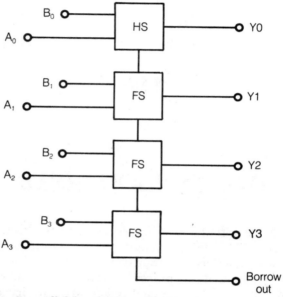

Fig. 9-12. A parallel four-bit binary subtractor.

10

Binary Counters

COUNTER circuits may be asynchronous or synchronous. The main difference is that with asynchronous counters all operations (except clear) are initiated by the incoming pulses, whereas with synchronous a separate clock pulse is employed to synchronize operations. Synchronous counter circuits are more complicated to design and generally use more components, but are usually faster in operation.

The basic element employed in a binary counter is a two-state (bistable) electrical device which is either off (0) or on (1), such as a flip-flop. A simple element of this type provides a count of 2^0 (decimal 1). The counting range can be extended by connecting a number of units in series, any overflow count from a preceding unit being an input to the following unit.

THE BASIC RIPPLE COUNTER

Indication of the state (position) of the count can be provided by tapping points showing the state of that stage. A further requirement is a means of resetting all stages to off (0), to clear the circuit after making a count via a clear signal. A four-stage counter as shown would then have a count capacity of $2^0 + 2^1 + 2^2 + 2^3 = 1 + 2 + 4 + 8 = 15$ decimal, although the actual number of combinations possible are $2^4 = 16$. The last pulse would produce overflow; that is, returning all four stages to 0 and carrying a 1 on to a fifth stage, if present. The count capacity of such a stage is therefore $2^n - 1$, where n is the number of flip-flop stages.

On the face of it, it would appear possible to use this spare pulse to clear a $2^n - 1$ counter circuit. This is so, except that the process would be tedious. To clear after a count, as many pulses would have to be applied to bring the count exactly to 2^n. Using a separate clear signal, all stages can be returned to 0 with a single pulse.

Such a form of cascaded circuit is generally known as a *ripple counter* because the changes in outputs of the flip-flops ripple through the counter from input to output. A basic circuit and the corresponding waveforms produced by a 4-bit ripple counter is shown in FIG. 10-1.

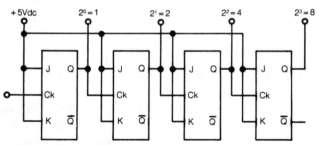

Fig. 10-1A. Basic arrangement for a four-stage binary counter.

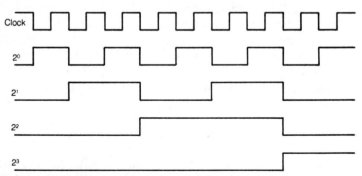

Fig. 10-1B. Waveform signals in a ripple counter.

In practice, unless all the flip-flops change state simultaneously, the waveforms may be spiked instead of square. It may therefore be necessary to treat the outputs in such a way that the counter is read only after these signals are stabilized. The other main limitation of the ripple counter is that ripple-through delays are cumulative and where many stages are involved, operating speed can be very slow. Such delays can be eliminated in a synchronous counter.

REVERSIBLE COUNTER

A *reversible counter* is designed to count either forwards or backwards, and is also known as an *up-down counter*. The Q output of the flip-flops is used for forward counting and the Q outputs for backward counting. The direction of counting is then determined by an up/down control signal X (such as $X = 1$ for up, $X = 0$ for down) applied to logic gates between the stages as shown in FIG. 10-2.

DECADE COUNTER

It is often desirable to have the counter circuit count to base 10 instead of 2; that is, in decimal rather than binary numbers. It is readily possible to utilize a ripple counter in this way, starting with the necessity of providing 10 combinations to cover a count of decimal 10. Again, the least number of flip-flop stages (bits) required to do this is four (giving $2^4 = 16$ possible combinations; $2^3 = 8$ would not be enough; and $2^5 = 32$ would be far more than necessary).

The basic circuit is shown in FIG. 10-3. The principle involved is that at a count of 10 (binary 1010), all binaries are reset to zero via a feedback line containing a NAND gate, the output from which feeds all clear inputs in parallel. At a count of 10, output states are:

$$Q_0 = 0 \quad Q_1 = 1 \quad Q_2 = 0 \quad Q_3 = 1$$

Inputs to the NAND gate are thus Q_1 and Q_3. After the tenth pulse Q_1 and Q_3 both go to 1, the output of the NAND

Decade Counter 137

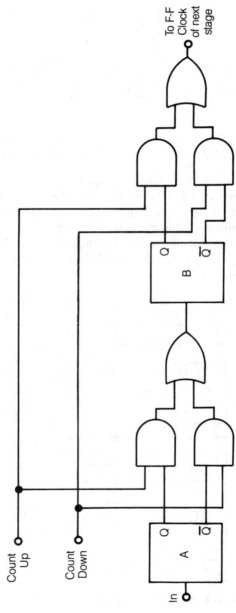

Fig. 10-2. Basic circuit for an up/down counter.

138 Binary Counters

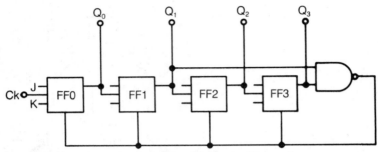

Fig. 10-3. Basic circuit for a decade (decimal) counter.

gate goes to 0, and FF0 and FF2 are reset to 0. Q_1 and Q_3 similarly return to 0 after a short delay. This delay, called a *propagation delay*, can be troublesome unless eliminated, so the feedback line normally incorporates a latching circuit to memorize and hold the output of the NAND gate until all flip-flops clear. A typical decade counter IC is shown in FIG. 10-4.

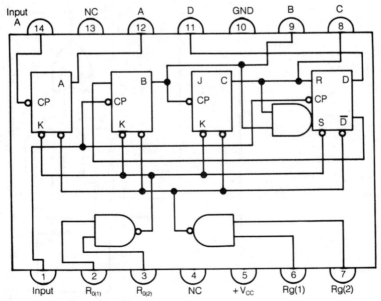

Fig. 10-4. A typical IC decade counter (7490).

To extend decimal counting beyond 10, it is necessary only to add further four-bit counters in cascade. Essentially then, to count to 100 (10^2), two decade counters in cascade; to count to 1000 (10^3), three decade counters in cascade; and so on.

DIVIDE-BY-N COUNTER

Exactly the same principle as used in the decade counter applies when designing a counter to count to any base N. The number of flip-flops required (n) is the smallest number for which $2^n > N$. Feedback via a NAND gate is then introduced to reset all binaries at the count of N, with each input to the NAND gate being an output from those flip-flops in state 1 at the count of N. For example, a divide-by-5 counter needs three flip-flops. At a count of N = 5 their outputs are:

$$Q_0 = 1 \quad Q_1 = 0 \quad Q_2 = 1$$

Hence, Q_0 and Q_2 are the inputs to the NAND gate. This type of counter, also known as a *modulus 5* or *mod 5 ripple counter* is shown in FIG. 10-5. The modulus, then, of a counter is the number of counting states that a counter has before it begins to repeat itself. A further example of modulus count-

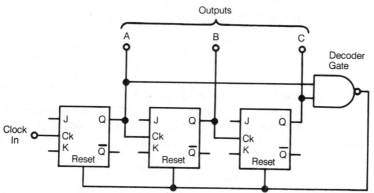

Fig. 10-5. A mod 5 ripple counter.

ing is a basic binary counter consisting of three flip-flops. This counter can count through eight discrete states ($2^3 = 8$) and is therefore said to have a *natural count* of 8. The same is true of a four-stage counter that can count through 16 discrete states ($2^4 = 16$). These counters are referred to as mod 8 and mod 16 counters.

Some divide-by-N counters are programmable, designed to accommodate a number of different N values, selectable at will. Basically, this involves having a suitable number of flip-flops to start with and selecting the N setting by connecting (or switching) the appropriate flip-flop outputs to the NAND gate inputs.

SYNCHRONOUS COUNTERS

In a synchronous counter circuit, all flip-flops are clocked simultaneously by the input pulses. Speed is thus limited only by the delay time of any one flip-flop, plus the propagation time of the control gate involved. In general terms, this usually makes them about twice as fast as ripple counters using similar components. There is also an absence of spikes in the output. These types of counters are known as *parallel counters*.

A typical basic circuit using T-type flip-flops is shown in FIG. 10-6. The requirement is that if $T = 0$ there is no change of state when the binary is clocked; and if $T = 1$ the flip-flop output is complemented with each pulse. In terms of T logic, this means:

$$T_0 = 1 \quad T_1 = Q_0 \quad T_2 = T_1 Q_1 \quad T_3 = T_2 Q_2$$

(Logic is performed by the AND gates.)

A critical factor is the minimum time between pulses (T_{min}) as this governs the maximum signal pulse frequency which can be applied. This is given by:

$$T_{min} = T_F + (n - 2) T_G$$

Where T_F is the propagation time of one flip-flop
 T_G is the propagation time of one AND gate
 n is the number of AND gates

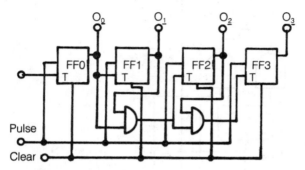

Fig. 10-6. Synchronous counter using T-type flip-flops.

Maximum signal points frequency is then equivalent to $1/T_{min}$.

Speed of operation can be improved by parallel rather than series working of the control gates, using a multiple-input AND gate taking inputs from every preceding flip-flop. This does, however, have the disadvantage of needing a large fan-in and fan-out, with heavier circuit loading. Nevertheless, parallel working is widely used, particularly for synchronous forward-backward counters and decade counters.

Synchronous Reversible Counter

A typical synchronous reversible (up/down) counter is shown in FIG. 10-7. Again control gates are interposed between the flip-flops but here they perform both up/down logic and (parallel) carry logic, simplifying the circuitry to some extent.

Synchronous Divide-by-N Counters

Design of synchronous counter circuits for decade counters or divide-by-N working can be extremely tedious,

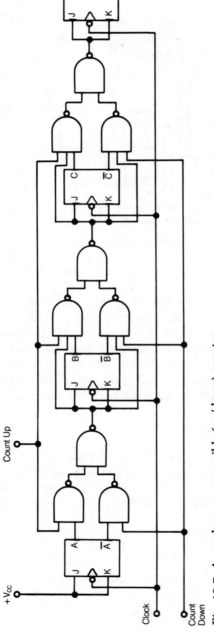

Fig. 10-7. A synchronous reversible (up/down) counter.

but can be simplified by the use of Karnaugh maps. These maps are used to simplify Boolean algebra (and thus simplify circuitry even further). Numerous examples, however, are available in IC form and are normally used in circuitry rather than start-from-scratch circuits. It is then only necessary to know the IC circuit characteristics and working parameters, and lead identification.

JOHNSON COUNTER (TWISTED RING COUNTER)

The circuit shown in FIG. 10-8 comprises five flip-flops connected with feedback from output to input, resulting in a continuous loop or ring being formed. Because the ring is crossed over or twisted at the input, it is known as a *twisted ring counter*. Alternatively, because it generates a *Johnson code* (a form of binary code) it is also called a *Johnson counter*. This counter is also called a *shift counter* since the waveforms literally shift through the flip-flops and the operation is cyclic in nature.

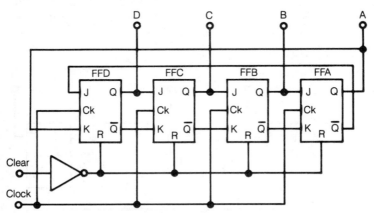

Fig. 10-8. A Johnson (twisted ring) counter.

The working principle is as follows. Starting with all outputs zero (A=0, B=0, etc.), after the first pulse the feedback loop applies the complement of A to FF4 and a 1 appears at E. Successive pulses shift this 1 along the counter so that after 5 pulses A=1, B=1, etc. The sixth pulse shifts $\overline{A}(0)$ into FF4 and succeeding pulses similarly up to the ninth pulse when A=1, B=0, C=0, D=0, and E=0. The tenth pulse then shifts a 0 into FF4 and all inputs are zero again. This counter, then sets each bit in a sequential order, beginning with the least significant bit (LSB) which is E or 2^0.

In effect, this circuit is a 1 to 10 (decimal) counter. In point of fact it has $2^5=32$ possible combinations, or the capacity to generate three different coded sequences of 10 decimal sequences.

IC BINARY COUNTERS

An example of an IC (7493 Binary Counter) providing a complete binary ripple through counter circuit is shown in FIG. 10-9. This is a TTL 14-pin device available in a dual-in-line package. This is a divide-by-2, divide-by-8 ripple counter, which, when externally connected will form a divide-by-16 counter. To reset the counter to 0, both reset-to-0 ($R_{0(1)}$ and $R_{0(2)}$) are taken to +5 volts (high). Either or both inputs to R_0 must be at ground for normal counting.

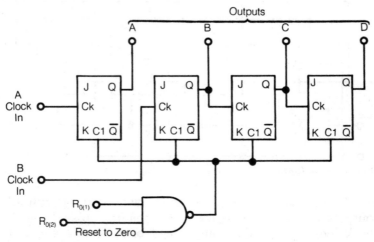

Fig. 10-9. The 7493 binary counter IC.

IC SYNCHRONOUS COUNTERS

FIGURE 10-10 shows a TTL synchronous up/down counter which this time is a 16 pin package (74193 4-bit binary up/down counter). It can count up from 0 to 15 and it can count down from 15 to 0. This has been a very popular counter because, besides being operated in the synchronous mode, the outputs may also be preset to any state simply by entering the required data at the data inputs while the load input is low (0). This allows the output to agree with what has been entered without being influenced by the count pulses. The advantage of this is that by changing the count length with the preset inputs, the counter can be used as a programmable divider.

The 74193 can also be cascaded without the need for external components. This makes possible counting numbers greater than 15 just by connecting the borrow and carry outputs of the first counter to the clock-down and clock-up inputs of the subsequent counter.

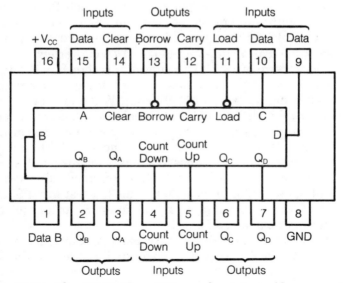

Fig. 10-10. The 74193 IC, a TTL synchronous up/down counter (four-bit).

11

Converters and Registers

DIGITAL-TO-ANALOG CONVERTERS (D/A)

In a digital circuit, data is represented by a series of digits, any change taking place in discrete steps. In many applications, it is desirable to be able to present this data in the form of a continuous steady voltage or circuit which then varies smoothly with any change of state (an analog signal infinitely variable between two limits). Systems for providing this are known as *digital-to-analog* or *D/A converters*.

A basic form of a 4-bit D/A converter is shown in FIG. 11-1 using a simple weighted resistor network. Input to each resistor is via a digital switch (S_0, S_1, etc.). When any switch is closed, or equivalent to an input signal of 1, a constant reference voltage (V_R) is applied through the corresponding resistor. Resistor values are chosen so that the signal outputs in each line have weighted values in a binary manner, 1, 2, 4, 8. Then a 1 at input S_0 gives an output of weighted value 1; a 1 at input S_1 an output of weighted value 2, a 1 at input S_2 an output of weighted value 4; a 1 at input S_3 an output of weighted value 8; and so on.

Put another way, since the same (constant) reference voltage is applied to each line when the input to that line is

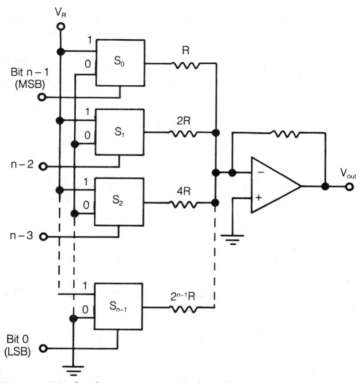

Fig. 11-1. Weighted resistor network digital-to-analog converter.

1, resistor values must be chosen so that:

- Line output voltage from S_3 is twice that in line from S_2
- Line output voltage from S_2 is twice that in line from S_1
- Line output voltage from S_1 is twice that in line from S_0

This effectively gives weights of 8, 4, 2, and 1 to the first four output lines, and so on. The total output voltage resulting from all lines is then fed to an op-amp to present the final output required as a current (the op-amp working as a voltage-to-current converter).

As an example, suppose the digital value is 1010 (decimal 10). Corresponding inputs are:

$$\begin{aligned} \text{to } S_3 &= 1 \\ \text{to } S_2 &= 0 \\ \text{to } S_1 &= 1 \\ \text{to } S_0 &= 0 \end{aligned}$$

If any input is 0 there is no output in that line (the digital switch remains open). Output in this case is therefore:

$$(1 \times 8) + (0 \times 4) + (1 \times 2) + (0 \times 1) = 10$$

That is to say, the 1 inputs at S_3 and S_1 give a final output of value 10 (the decimal equivalent). The same principle can be extended to cover any number of bits. Thus for an N-bit D/A converter the following general relationship applies:

$$V_{out} = V_R(B_{n-1}2^{-1} + B_{n-2}2^{-2} + B_{n-3}2^{-3} \ldots B_0 2^{-n})$$

B_n represents the binary word. This defines the weighting necessary. The most significant bit (B_{n-1}) has a weight of $V_R/2$, down to the least significant bit (B_0) which has a weight of $V_R/2n$. Thus with a 6-bit converter, for example, the equation becomes:

$$V_{out} = V_R/64(32n_5 + 16n_4 + 8n_3 + 4n_2 + 2n_1 + n_0)$$

The basic disadvantages of such a circuit are that it demands stable, close tolerance resistors with values extending over a wide range, the output resistance can be quite high, and the output signal is not a convenient multiple of the digital input value. Other circuits are therefore normally preferred in practice. One of these is the serial converter which works as an integrator, or a ladder type circuit.

The ladder type D/A converter is more complex in that it requires twice the number of resistors to handle the same number of bits, but these need only be of two values, R and 2R. Actual resistor values are not as important as the correct 1:2 ratio values. A basic circuit of this type is shown in FIG. 11-2. Here the necessary weighting of signals is achieved by current splitting. At the top of any ladder the current splits equally right and left, yielding weightings corresponding to $V_R/2$, $V_R/4$, $V_R/8$...down to $V_R/2^n$.

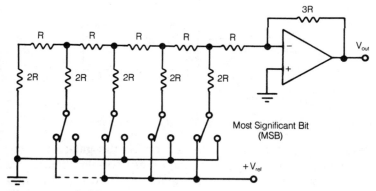

Fig. 11-2. A ladder-type D/A converter.

ANALOG-TO-DIGITAL CONVERTERS (A/D)

An analog-to-digital (A/D) converter converts the infinitely variable analog data signals into digital form. There are many forms of such devices, but the main types are voltage-to-frequency converters, pulse counters, and integrating converters.

Voltage-to-frequency converters are based on a voltage-controlled oscillator where the output is applied to a counter for a period of time controlled by a clock pulse generator. Since this output frequency is proportional to input voltage, the counter can be calibrated to read out the digital equivalent to the analog input.

A basic example of a counter type circuit is shown in FIG. 11-3. When an analog signal (V_s) is applied to the com-

150 Converters and Registers

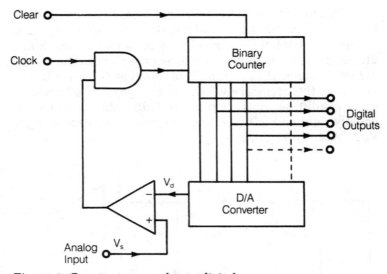

Fig. 11-3. Counter-type analog-to-digital converter.

parator there is an output which opens the gate, allowing clock pulses to be applied to the binary counter. The count continues until the feedback signal (V_d) from the D/A converter becomes equal to V_s, when the comparator output falls to zero and the count is frozen in the binary counter and displayed or read out. In other words, the count proceeds one step at a time until a final balance is reached. For example, to establish a count of 9.9 in 0.1 steps would involve 99 pulses passing through the gate before a final balance is reached; or 999 pulses to count up to 99.9 with the same interval, and so on. The speed of conversion thus depends both on the pulse rate and the method by which final balance is obtained.

A more rapid method of conversion is possible using successive approximations. Here the first clock pulse sets the counter to one-half of the maximum output. The next pulse then sets the counter to one-half of a half in a plus or minus manner; that is, plus if V_s is greater than V_d and minus if V_d is greater than V_s, and so on with following pulses. This enables the final balance to be reached more quickly.

SHIFT REGISTERS

A digital memory device has a one-bit capacity, so to store or register an N-bit word, N memory units (flip-flops) are required. It is then necessary to cascade the flip-flops output-to-output to feed input data into the system serially. It is this facility to shift the data along the circuit that gives such a device the name *shift* register.

A basic circuit for performing this function is shown in FIG. 11-4. Each flip-flop is a master-slave type, the stage used to store the most significant bit (MSB) having S and R terminals connected together via an inverter to turn it into a D type latch. Starting with all outputs clear ($Q_0 = 0$, $Q_1 = 0$, etc.) Cr is set to 1 and Pr held at 1 by keeping preset enable at 0. Clock pulses are now applied. The first pulse (corresponding to the least significant bit) enters FF4 which latches, changing Ck from 0 to 1. Output Q_4 is now at 1 with all other outputs at 0.

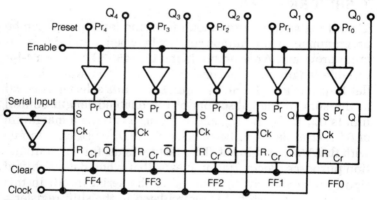

Fig. 11-4. *Basic circuit for a five-bit shift register.*

Each succeeding pulse then shifts the preceding pulse(s) to the right to make room for the incoming digit until after five pulses (or N pulses in an N-bit register), the full input word has been taken into the register. At that point the input pulses must stop. This sequence of operations can be seen from the following diagram, taking as an example 10110 as the 5-bit word fed into a 5-bit shift register.

Clock Pulse	Word Bit	MSB				LSB
		Q_4	Q_3	Q_2	Q_1	Q_0
1	1 ⟶	1	0	0	0	0
2	0 ⟶	0	1	0	0	0
3	1 ⟶	1	0	1	0	0
4	1 ⟶	1	1	0	1	0
5	0 ⟶	0	1	1	0	1

Such a shift register accepts input serially and gives a parallel output, and so is properly described as a series-in, parallel-out register (SIPO). Other modes of working are possible:

- Series-in, series-out (SISO)
- Parallel-in, parallel-out (PIPO)
- Parallel-in, series-out (PISO)

IC SHIFT REGISTER

IC shift registers are produced in varying lengths and can be programmed to any number of bits between 1 and the maximum provided. An example is the HEF4557B 1-bit to 64-bit variable-length shift register in FIG. 11-5. It is available as a flat 16-pin DIP with LSI. The number of bits selected is equal to the sum of the subscripts of the enabled length control inputs (L_1, L_2, L_4, L_8, L_{16}, and L_{32}) plus 1, giving a maximum of 64. Serial data can be selected from D_A or D_B data inputs with the A/$\overline{B}$ select input. This feature is useful for recirculation purposes. *Recirculation* means that when data is shifted right, the MSB may be returned to the serial input. In this way data is not lost, but is recirculated in the shift register.

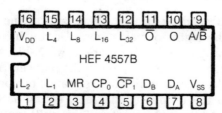

Fig. 11-5. *IC 1-to-64-bit shift register (HEF4557B).*

Information on D_A or D_B is shifted one position to the right on the LOW to HIGH transition of CP_0 while $\overline{CP}_1$ is LOW; or on the HIGH to LOW transition of $\overline{CP}_1$ while CP_0 is HIGH. When HIGH, master reset (MR) resets the whole register asynchronously (0 = LOW; $\overline{0}$ = HIGH) and independent of the other inputs. The complete logic diagram is shown in FIG. 11-6.

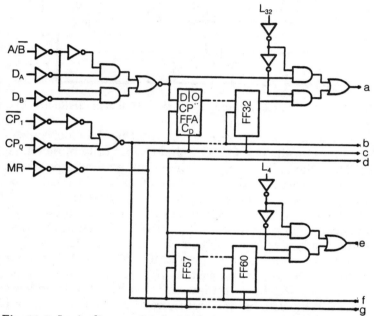

Fig. 11-6. Logic diagram for the IC shift register of Fig. 11-5.

This device can work on any voltage from 5-15V, drawing a quiescent current of 50-200μA. Propagation delay is on the order of 240-260 ns, depending on voltage. The maximum clock pulse frequency is 5 MHz with a 5V supply and up to 20 MHz with a 15V supply.

Another example of the logic provided by an IC shift register circuit is shown in FIG. 11-7. In effect, this is a serial-to-parallel converter. Information present on the data input D_A is shifted to the first register position and all the data in the reg-

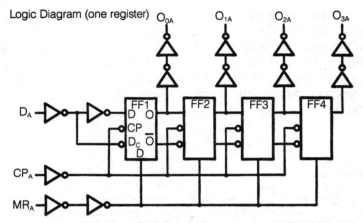

Fig. 11-7. IC series-to-parallel converter shift register (HEF4058B).

ister is shifted one position to the right by the clock pulse. The four outputs O_{0A}, O_{1A}, O_{2A}, and O_{A3} are fully buffered. A HIGH (1) signal on the asynchronous master reset input (MR) clears the register and returns O_0 to O_3 to LOW (0), irrespective of the clock input and the serial data input (D_A).

Another IC package (Signetics HEF4015B) actually contains two such systems in a 16-pin DIP as shown in FIG. 11-8. Additionally, CMOS integrated circuitry lends itself well to high component density. Some CMOS IC shift registers are available with 64 stages, each stage configured as a D type master-slave flip-flop. In these shift registers the logic level present at the data input is transferred into the first stage and

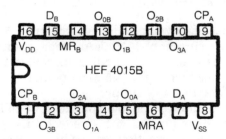

Fig. 11-8. IC package providing two shift registers.

shifted one stage at each positive-going clock transition. FIGURE 11-9 is an illustration of the IC package of this type of shift register. As you can see, it is an HBC4031A. Also shown is its logic diagram.

In this device, information can be permanently shared with the clock line in either the LOW or HIGH state. There is also a mode input control which allows operation in the recirculating mode when in the HIGH state. Register packages can be cascaded and clock lines driven directly for fast working, or, alternately, a delayed clock output is provided allowing reduced clock drive fan-out and transition time when cascaded. The entire circuitry is contained in a 16-pin DIP package.

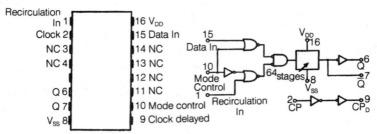

Fig. 11-9. CMOS HBC4031A IC with logic diagram.

DYNAMIC MOS SHIFT REGISTER

A basic circuit for a dynamic MOS shift register stage is shown in FIG. 11-10. It employs two separate clock inputs; that is, it is a two-phase MOS system, each stage incorporating six MOSFETs. Specifically, these provide two NAND gates in cascade, each clock pulse shifting and inverting a bit through that stage. In this device, a minimum clock rate is essential to retain gate capacitance (necessary for retaining memory) and a maximum clock is also essential, limited, however, by the response rate of the circuit. It is a general feature of most IC dynamic MOS shift registers that both input and output are compatible with TTL integrated circuits.

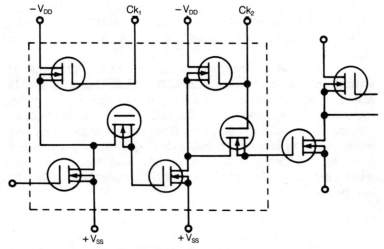

Fig. 11-10. Dynamic MOS shift register.

12

The Arithmetic Logic Unit (ALU)

THE arithmetic logic unit (ALU) of a computer is that part of the circuitry that can perform arithmetic and logic functions on data. Typically, this device can add, subtract, AND, OR, XOR, complement, shift right, shift left, increment, and decrement, and is part of a larger circuit called the MPU, or microprocessor unit. Much of the material that you have read thus far describes devices that, when arranged in an appropriate manner, serve as the building blocks of the ALU; for example, the half and full-adders and shift registers.

Typically, ALUs provide four-bit arithmetic operations with up to 16 instruction capability. The 74181, a four-bit MSI TTL ALU actually contains 32 separate operations and divides these operations into arithmetic and logic functions. This chapter discusses the 74181 and its two modes of operations, arithmetic and logic. Additionally, ALUs may be cascaded to increase word size without affecting the operation of individual functions.

CASCADING ALUs

ALUs can be cascaded by using the 74182, a look-ahead carry generator. As shown in FIG. 12-1, this device allows parallel operations of carry and borrow functions for two ALUs

158 The Arithmetic Logic Unit (ALU)

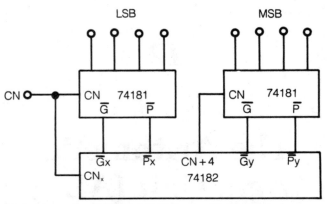

Fig. 12-1. Cascading 74181s (ALUs) using the 74182.

(four ALUs are possible with the 74182). Parallel transfer of carry and borrow from one ALU to another is achieved by the use of additional outputs from the 74181 called the *carry generate* (G) and the *carry propagate* (P).

Another method is to apply the CN+4 (ripple carry) output of the least significant four-bit word of the least significant ALU into the CN (ripple carry) input of the following ALU.

ALU FUNCTIONS

TABLE 12-1 is a listing of all the functions of the 74181 ALU. As you can see, there are 16 arithmetic functions and 16 logic functions. Positive logic operation is performed when CN is high. Negative logic is used when CN is low. This is for the logic mode of operation only.

Arithmetic Operation

To implement arithmetic operations in the following examples, M must be low (0) and CN must be low (0). Each condition of the inputs, as shown in TABLE 12-1, is stated as well as the output results for those input conditions. Also, assume A = 1001 and B = 0110.

Table 12-1. 74181 Function Table

Mode Select Inputs				Active High Inputs & Outputs	
				Logic	Arithmetic
S_3	S_2	S_1	S_0	M = 1	M = 0, CN = 0
0	0	0	0	$\overline{A}$	A
0	0	0	1	$\overline{A + B}$	A + B
0	0	1	0	$\overline{A}B$	$A + \overline{B}$
0	0	1	1	Logic 0	minus 1
0	1	0	0	$\overline{AB}$	A plus $A\overline{B}$
0	1	0	1	$\overline{B}$	(A + B) plus $A\overline{B}$
0	1	1	0	A ⊕ B	A minus B minus 1
0	1	1	1	$A\overline{B}$	AB minus 1
1	0	0	0	$\overline{A} + B$	A plus AB
1	0	0	1	$\overline{A \oplus B}$	A plus B
1	0	1	0	B	$(A + \overline{B})$ plus AB
1	0	1	1	AB	AB minus 1
1	1	0	0	Logic 1	A plus A
1	1	0	1	$A + \overline{B}$	(A + B) plus A
1	1	1	0	A + B	$(A + \overline{B})$ plus A
1	1	1	1	A	A minus 1

Condition 1:
   Selection = 0000 = A
   $F_0 - F_3$ = output = A = 1001
   CN+4 = 1(no carry)

Condition 2:
   Selection = 0001 = A+B
   $F_0 - F_3$ = output = 1001 + 0110 = 1111
   CN+4 = 1(no carry)

Condition 3:
   Selection = 0010 = A + $\overline{B}$
   $F_0 - F_3$ = output = 1001 + $\overline{0110}$ = 1001
   CN+4 = 1(no carry)

Condition 4:
   Selection = 0011 = minus 1(2's complement)
   $F_0 - F_3$ = output = 1111
   CN+4 = 1(borrow)

The Arithmetic Logic Unit (ALU)

Condition 5:
    Selection = 0100 = A Plus A$\overline{B}$
    $F_0 - F_3$ = output = 1001 Plus 1001($\overline{0110}$) = 0010
    CN+4 = 0(carry)

Condition 6:
    Selection = 0101 = (A+B) Plus A$\overline{B}$
    $F_0 - F_3$ = output = (1001+0110) Plus 1001($\overline{0110}$) = 1000
    CN+4 = 0(carry)

Condition 7:
    Selection = 0110 = A Minus B Minus 1
    $F_0 - F_3$ = output = 1001 Minus 0110 Minus 1 = 0010
    CN+4 = 0(no borrow)

Condition 8:
    Selection = 0111 = A$\overline{B}$ Minus 1
    $F_0 - F_3$ = output = 1001($\overline{0110}$) Minus 1 = 1000
    CN+4 = 0(no borrow)

Condition 9:
    Selection = 1000 = A Plus AB
    $F_0 - F_3$ = output = 1001 Plus 1001(0110)
    CN+4 = 1(no carry)

Condition 10:
    Selection = 1001 = A Plus B
    $F_0 - F_3$ = output = 1001 Plus 0110 = 1111
    CN+4 = 1(no carry)

Condition 11:
    Selection = 1010 = (A+$\overline{B}$) Plus AB
    $F_0 - F_3$ = output = 1001+$\overline{0110}$ Plus 1001(0110) = 1001
    CN+4 = 1(no carry)

Condition 12:
    Selection = 1001 = AB Minus 1
    $F_0 - F_3$ = output = 1001(0110) Minus 1 = 1111
    CN+4 = 1(borrow)

Condition 13:
   Selection = 1100 = A Plus A
   $F_0 - F_3$ = output = 1001 Plus 1001 = 0010
   CN+4 = 0(carry)

Condition 14:
   Selection = 1101 = (A+B) Plus A
   $F_0 - F_3$ = output = (1001 + 0110) Plus 1001 = 1000
   CN+4 = 0(carry)

Condition 15:
   Selection = 1110 = (A+$\overline{B}$) Plus A
   $F_0 - F_3$ = output = (1001 + $\overline{0110}$) Plus 1001 = 0010
   CN+4 = 0(carry)

Condition 16:
   Selection = 1111 = A Minus 1
   $F_0 - F_3$ = output = 1001 Minus 1 = 1000
   CN+4 = 0(no borrow)

Logic Operation

In this mode of operation M=1 and CN=1. A=1001 and B=0110 in this operation also. In the logic mode of operation, CN+4 is not a valid output and therefore you will not see this listed as a resultant output. That particular output is valid only in the arithmetic mode of operation.

Condition 1:
   Selection = 0000 = $\overline{A}$
   $F_0 - F_3$ = output = $\overline{1001}$ = 0110

Condition 2:
   Selection = 0001 = $\overline{A+B}$
   $F_0 - F_3$ = output = $\overline{1001+0110}$ = 0000

Condition 3:
   Selection = 0010 = $\overline{A}B$
   $F_0 - F_3$ = output = $\overline{1001}$(0110) = 0110

162 The Arithmetic Logic Unit (ALU)

Condition 4:
  Selection = 0011 = Logic 0
  $F_0 - F_3$ = output = 0000 (Inputs A and B have no effect on the output; it is always logic 0 with a selection input of 0011.)

Condition 5:
  Selection = 0100 = $\overline{A}B$
  $F_0 - F_3$ = output = $\overline{1001}(0110)$ = 1111

Condition 6:
  Selection = 0101 = $\overline{B}$
  $F_0 - F_3$ = output = $\overline{0110}$ = 1001

Condition 7:
  Selection = 0110 = $A \oplus B$
  $F_0 - F_3$ = output = 1001 $\oplus$ 0110 = 1111

Condition 8:
  Selection = 0110 = $A\overline{B}$
  $F_0 - F_3$ = output = 1001($\overline{0110}$) = 1001

Condition 9:
  Selection = 1000 = $\overline{A}+B$
  $F_0 - F_3$ = output = $\overline{1001}$ + 0110 = 0110

Condition 10:
  Selection = 1001 = $\overline{A \oplus B}$
  $F_0 - F_3$ = output = $\overline{1001 \oplus 0110}$ = 0000

Condition 11:
  Selection = 1010 = B
  $F_0 - F_3$ = output = 0110

Condition 12:
  Selection = 1011 = AB
  $F_0 - F_3$ = output = 1001(0110) = 0000

Condition 13:
   Selection = 1100 = Logic 1
   $F_0 - F_3$ = output = 1111   (Inputs A and B have no effect on the output; it is always logic 1 with a selection input of 1100.)

Condition 14:
   Selection = 1101 = $A + \overline{B}$
   $F_0 - F_3$ = output = 1001 + $\overline{0110}$ = 1001

Condition 15:
   Selection = 1110 = $A + B$
   $F_0 - F_3$ = output = 1001 + 0110 = 1111

Condition 16:
   Selection = 1111 = A
   $F_0 - F_3$ = output = 1001

The logic mode of operation and the arithmetic mode of operation together make up 32 individual operations for the 74181 ALU. The ALU is just one part, although a very important part, of a microprocessor unit (MPU). The MPU is, in effect, the brains of the computer. It interprets instructions stored in memory, step by step (sequentially), and manipulates that set of instructions (data) to perform a certain task.

MICROPROCESSORS

A microprocessor unit (MPU) is a device, usually a single integrated circuit, that acts on data to perform a single task. It was stated earlier in this chapter that the ALU is part of the MPU and that the MPU is the brain or thinking part of the computer. Actually, the MPU itself is not capable of performing tasks on its own. It must interpret information, usually a set of instructions called a program, and then be allowed to control other devices.

A popular microprocessor in use is the Motorola 6802. It's been around for a number of years. It is a member of the 6800 family of micros and contains 16 unidirectional address lines with 8 bidirectional data lines. For clarity, *address lines* make up the *address bus* which determines which device or memory location is accessed. When information is accessed, a signal is sent down the address lines to the correct memory location. In computers, data is stored in memory locations or *cells* and each cell has an address. As for data lines, they make up the data bus which normally acts as an input to the system. In the MPU the data bus acts as an output. FIGURE 12-2 is a block diagram of the 6802 microprocessor.

When data comes down the data bus (usually as two eight-bit words) and into the MPU, it is manipulated by the ALU. The ALU then supplies an eight-bit answer that is placed into either accumulator A or B. Additionally, six output test bits are applied to the condition code register (CCR). These six test bits determine the results of the instruction performed or special conditions in reference to the two eight-bit words.

A description of the blocks within the MPU may be helpful here.

Temporary A & B Data Registers. These registers provide storage capability for the ALU. These are eight-bit data registers and are written into by the A & B accumulators.

Accumulator A & B (Acc A & Acc B). These are also eight-bit registers used to transfer data into and out of the ALU via the data bus. Initially, data is transferred into one of the accumulators. Once manipulation on the data has taken place by the ALU, the result is stored in one of the accumulators.

Instruction Code Register. This register holds the eight-bit instruction that is being performed. Input comes in from the data bus and the instruction decoder receives its output. During the operation of a program, this register is loaded first.

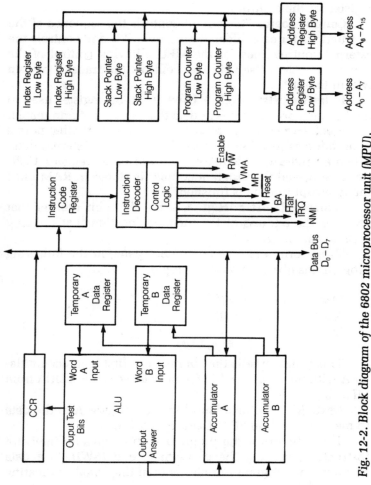

Fig. 12-2. Block diagram of the 6802 microprocessor unit (MPU).

Instruction Decoder. This circuit produces a logic code from the eight-bit instruction code of the instruction code register for the control logic.

Control Logic. This circuit controls the sequence of the instructions for each block within the MPU and controls the external control lines for the MPU allowing control of the computer system itself. It also controls the transfer of data within the MPU.

Index Register (High & Low). A 16-bit register used to modify memory locations or data that is user programmable.

Stack Pointer (High & Low). A 16-bit register that points to an address location in RAM for storage of internal data. When an interrupt occurs, data in the index register, CCR, Acc A & B, and the program counter is stored in RAM until the interrupt routine is finished.

Address Register (High & Low). A 16-bit register that contains the address of the memory location that is being accessed at any given time.

To perform a simple computer operation the following program may be used as an example:

LDAA from MLoc 6F
ADDA from MLoc 5E
SWI

This program tells the computer to first load accumulator A with the data found in memory location 6F (LDDA from MLoc 6F).

Next, ADD the data found in accumulator A to the data in memory location 5E (ADDA from MLoc 5E).

Then, interrupt the program with a software interrupt instruction, causing the program to end (SWI). The data found in the Program Counter is then displayed, indicating the completion of the program.

The computer program just listed used mnemonics (a group of letters that symbolize an instruction) as a list of instructions, but computers need binary code to understand what to do. The following program, therefore, is written in hexadecimal and is an exact replica of the program above:

 $96 $6F
 $9B $5E
 $3F

(The dollar sign, $, represents hexadecimal.)

Along with the 6800 series of microprocessors, Motorola also has a 68000 series of microprocessors. In addition, this particular manufacturer has now come up with a new microprocessor chip designated the 88000, which is a reduced instruction set computer (RISC).

Appendix A

Binary/Decimal Equivalents

Decimal	2^5	2^4	2^3	2^2	2^1	2^0
	(32)	(16)	(8)	(4)	(2)	(1)
0						0
1						1
2						1	0
3						1	1
4					1	0	0
5					1	0	1
6					1	1	0
7					1	1	1
8				1	0	0	0
9				1	0	0	1
10				1	0	1	0
11				1	0	1	1
12				1	1	0	0
13				1	1	0	1
14				1	1	1	0
15				1	1	1	1
16		1	0	0	0	0
17		1	0	0	0	1
18		1	0	0	1	0
19		1	0	0	1	1
20		1	0	1	0	0
21		1	0	1	0	1
22		1	0	1	1	0
23		1	0	1	1	1
24		1	1	0	0	0

Binary/Decimal Equivalents

Decimal	2^5	2^4	2^3	2^2	2^1	2^0
	(32)	(16)	(8)	(4)	(2)	(1)
25			1	1	0	0	1
26			1	1	0	1	0
27			1	1	0	1	1
28			1	1	1	0	0
29			1	1	1	0	1
30			1	1	1	1	0
31			1	1	1	1	1
32		1	0	0	0	0	0
33		1	0	0	0	0	1
34		1	0	0	0	1	0
35		1	0	0	0	1	1
36		1	0	0	1	0	0
37		1	0	0	1	0	1
38		1	0	0	1	1	0
39		1	0	0	1	1	1
40		1	0	1	0	0	0
41		1	0	1	0	0	1
42		1	0	1	0	1	0
43		1	0	1	0	1	1
44		1	0	1	1	0	0
45		1	0	1	1	0	1
46		1	0	1	1	1	0
47		1	0	1	1	1	1
48		1	1	0	0	0	0
49		1	1	0	0	0	1
50		1	1	0	0	1	0
51		1	1	0	0	1	1
52		1	1	0	1	0	0
53		1	1	0	1	0	1
54		1	1	0	1	1	0
55		1	1	0	1	1	1
56		1	1	1	0	0	0
57		1	1	1	0	0	1
58		1	1	1	0	1	0
59		1	1	1	0	1	1
60		1	1	1	1	0	0
61		1	1	1	1	0	1
62		1	1	1	1	1	0
63		1	1	1	1	1	1
64	1	0	0	0	0	0	0
65	1	0	0	0	0	0	1
66	1	0	0	0	0	1	0

170 Binary/Decimal Equivalents

Decimal	2^6	2^5	2^4	2^3	2^2	2^1	2^0
	(64)	(32)	(16)	(8)	(4)	(2)	(1)
67	1	0	0	0	0	1	1
68	1	0	0	0	1	0	0
69	1	0	0	0	1	0	1
70	1	0	0	0	1	1	0
71	1	0	0	0	1	1	1
72	1	0	0	1	0	0	0
73	1	0	0	1	0	0	1
74	1	0	0	1	0	1	0
75	1	0	0	1	0	1	1
76	1	0	0	1	1	0	0
77	1	0	0	1	1	0	1
78	1	0	0	1	1	1	0
79	1	0	0	1	1	1	1
80	1	0	1	0	0	0	0
81	1	0	1	0	0	0	1
82	1	0	1	0	0	1	0
83	1	0	1	0	0	1	1
84	1	0	1	0	1	0	0
85	1	0	1	0	1	0	1
86	1	0	1	0	1	1	0
87	1	0	1	0	1	1	1
88	1	0	1	1	0	0	0
89	1	0	1	1	0	0	1
90	1	0	1	1	0	1	0
91	1	0	1	1	0	1	1
92	1	0	1	1	1	0	0
93	1	0	1	1	1	0	1
94	1	0	1	1	1	1	0
95	1	0	1	1	1	1	1
96	1	1	0	0	0	0	0
97	1	1	0	0	0	0	1
98	1	1	0	0	0	1	0
99	1	1	0	0	0	1	1
100	1	1	0	0	1	0	0
101	1	1	0	0	1	0	1
102	1	1	0	0	1	1	0
103	1	1	0	0	1	1	1
104	1	1	0	1	0	0	0
105	1	1	0	1	0	0	1
106	1	1	0	1	0	1	0
107	1	1	0	1	0	1	1
108	1	1	0	1	1	0	0
109	1	1	0	1	0	0	1

Decimal	2^7	2^6	2^5	2^4	2^3	2^2	2^1	2^0
	(128)	(64)	(32)	(16)	(8)	(4)	(2)	(1)
110		1	1	0	1	1	1	0
111		1	1	0	1	1	1	1
112		1	1	1	0	0	0	0
113		1	1	1	0	0	0	1
114		1	1	1	0	0	1	0
115		1	1	1	0	0	1	1
116		1	1	1	0	1	0	0
117		1	1	1	0	1	0	1
118		1	1	1	0	1	1	0
119		1	1	1	0	1	1	1
120		1	1	1	1	0	0	0
121		1	1	1	1	0	0	1
122		1	1	1	1	0	1	0
123		1	1	1	1	0	1	1
124		1	1	1	1	1	0	0
125		1	1	1	1	1	0	1
126		1	1	1	1	1	1	0
127		1	1	1	1	1	1	1
128	1	0	0	0	0	0	0	0
129	1	0	0	0	0	0	0	1
130	1	0	0	0	0	0	1	0
131	1	0	0	0	0	0	1	1
132	1	0	0	0	0	1	0	0
133	1	0	0	0	0	1	0	1
134	1	0	0	0	0	1	1	0
135	1	0	0	0	0	1	1	1
136	1	0	0	0	1	0	0	0
137	1	0	0	0	1	0	0	1
138	1	0	0	0	1	0	1	0
139	1	0	0	0	1	0	1	1
140	1	0	0	0	1	1	0	0
141	1	0	0	0	1	1	0	1
142	1	0	0	0	1	1	1	0
143	1	0	0	0	1	1	1	1
144	1	0	0	1	0	0	0	0
145	1	0	0	1	0	0	0	1
146	1	0	0	1	0	0	1	0
147	1	0	0	1	0	0	1	1
148	1	0	0	1	0	1	0	0
149	1	0	0	1	0	1	0	1
150	1	0	0	1	0	1	1	0
151	1	0	0	1	0	1	1	1
152	1	0	0	1	1	0	0	0

Decimal	2^7	2^6	2^5	2^4	2^3	2^2	2^1	2^0
	(128)	(64)	(32)	(16)	(8)	(4)	(2)	(1)
153	1	0	0	1	1	0	0	1
154	1	0	0	1	1	0	1	0
155	1	0	0	1	1	0	1	1
156	1	0	0	1	1	1	0	0
157	1	0	0	1	1	1	0	1
158	1	0	0	1	1	1	1	0
159	1	0	0	1	1	1	1	1
160	1	0	1	0	0	0	0	0
up to								
255	1	1	1	1	1	1	1	1

and so on

Higher orders of binary number equivalents:

2^8	256	2^{30}	1073741824
2^9	512	2^{31}	2147483648
2^{10}	1024	2^{32}	4294967296
2^{11}	2048	2^{33}	8589934592
2^{12}	4096	2^{34}	17179869184
2^{13}	8192	2^{35}	34359738368
2^{14}	16384	2^{36}	68719476736
2^{15}	32768	2^{37}	137438953472
2^{16}	65536	2^{38}	274877906944
2^{17}	131072	2^{39}	549755813888
2^{18}	262144	2^{40}	1099511627776
2^{19}	524288	2^{41}	2199023255552
2^{20}	1048576	2^{42}	4398046511104
2^{21}	2097152	2^{43}	8796093022208
2^{22}	4194304	2^{44}	17592186044416
2^{23}	8388608	2^{45}	35184372088832
2^{24}	16777216	2^{46}	70368744177664
2^{25}	33554432	2^{47}	140737488355328
2^{26}	67108864	2^{48}	281474976710656
2^{27}	134217728	2^{49}	562949953421312
2^{28}	268435456	2^{50}	1125899906842624
2^{29}	536870912		

Appendix B

Simplifying Digital Logic Circuitry

TRUTH TABLES AND BOOLEAN ALGEBRA

Truth tables, Boolean Algebra, and minimizing are all used to represent logic circuitry in its simplest form. As an example, suppose the problem in simplifying a logic circuit involves three inputs (A, B, and C) and the logic to be provided is that there is an output with the following combinations of signals:

B AND C OR A AND C OR A AND B

The corresponding truth table can be written:

	A	B	C	S
Line 1	0	0	0	0
Line 2	0	0	1	0
Line 3	0	1	0	0
Line 4	0	1	1	1
Line 5	1	0	0	0
Line 6	1	0	1	1
Line 7	1	1	0	1
Line 8	1	1	1	1

This describes all of the states, twenty four possible combinations, of which only four provide one output. This may not be so obvious from the original statement, where it may appear that only three states provide an output. The solution for S can also be written in the form of a Boolean equation as such:

$$S = 1 = \overline{A}BC + A\overline{B}C + AB\overline{C} + ABC$$

Again, this indicates four states giving an output where S=1. These states can be provided by covering all states as laid down by the original logic statement, reminded by the truth table, and/or the Boolean equation that there are four possible states involved where S=1. Logic elements arranged in the combination shown in FIG. B-1 would then cover all states with what can be referred to as unreduced or unminimized combinations. A second look with a view to reducing or minimizing the actual combinations required can then be very worthwhile.

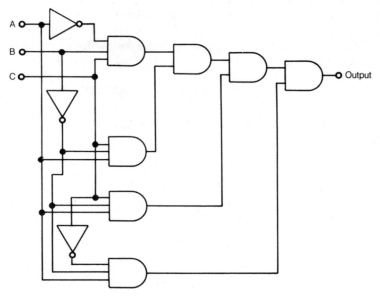

Fig. B-1. Unminimized logic element combination.

MINIMIZING

The Boolean equation above then simplifies to:

$$S = 1 = BC + AC + AB$$

This is merely the original statement expressed in Boolean algebra. The same follows from a study of the truth table. We only need the states established by the fourth, sixth, and seventh lines. The minimized circuit is then very much simpler, reducing the number of logic elements actually required from ten to five as shown in FIG. B-2.

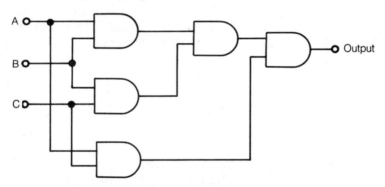

Fig. B-2. *Minimized logic element circuit.*

Minimizing is thus a very important part of logic circuit design. It can eliminate redundant or unnecessary components. It is not always easy to spot how this can be achieved working with block logic diagrams, but reducing the Boolean equation to its simplest form provides a positive answer, provided you do the Boolean algebra correctly. It is not so easy with combination circuits where minimizing is best done with the aid of a Karnaugh map (to be discussed later). The full design procedure then is:

1. Put down all input combinations which provide an output.

2. Construct a truth table as a check that all possible combinations have been considered.
3. Derive the Boolean equation which provides an output.
4. Construct a Karnaugh map for all participating variables.
5. Use this map to minimize the Boolean equation.
6. Draw up a circuit from this minimized equation.

MINTERMS AND MAXTERMS

Truth tables and Boolean algebra equations are closely related. In most digital circuit designs the starting point is the truth table from which the corresponding formula is derived. For example, here is the truth table for an XOR logic:

A	B	S
0	0	0
0	1	1
1	0	1
1	1	0

The corresponding equation is $\overline{A}B + A\overline{B} = S$

This particular equation is a sum of products, or what is called the normal minterm form when referring to switching circuits. A complementary formula can be devised for the same conditions (from the same truth table) by considering combinations which do not produce an output. This is called a dual equation and in this case is:

$$\overline{A}\,\overline{B} + AB = \overline{S}$$
$$\text{by inversion } S = \overline{\overline{A}\,\overline{B} + AB}$$
$$= (A + B)(\overline{A} + B)$$

This is the product of sums and this form of equation is called the *normal maxterm form*.

Specifically then, the minterm form of an equation, being a sum of products, can be solved by digital devices having an AND function. Maxterm forms of an equation, being a product of sums, can be solved by OR devices. The value of this is that a switching function requirement can be written in equations in either minterm for solution with AND devices, or maxterm form for solution with OR devices, and the two alternatives compared term for term.

Again, minterms and maxterms can be directly related to a truth table. For example, possible minterms and maxterms covering three binary variables A, B, and C are:

A	B	C	Minterm	Maxterm
0	0	0	$\overline{A}\,\overline{B}\,\overline{C}$	$\overline{A}+\overline{B}+\overline{C}$
0	0	1	$\overline{A}\,\overline{B}C$	$\overline{A}+\overline{B}+C$
0	1	0	$\overline{A}B\overline{C}$	$\overline{A}+B+\overline{C}$
0	1	1	$\overline{A}BC$	$\overline{A}+B+C$
1	0	0	$A\overline{B}\,\overline{C}$	$A+\overline{B}+\overline{C}$
1	0	1	$A\overline{B}C$	$A+\overline{B}+C$
1	1	0	$AB\overline{C}$	$A+B+\overline{C}$
1	1	1	ABC	$A+B+C$

Minterms and maxterms can also be devised directly from any functional expression f(A,B) where f is a Boolean function of the binary values A and B. As an example:

Minterm form = f(A,B)
= $\overline{A}\,\overline{B}f(0,0) + \overline{A}Bf(0,1) + A\overline{B}f(1,0) + ABf(1,1)$

Maxterm form = f(A,B)
= $(\overline{A}+B+f(1,1))(\overline{A}+B+f(1,0))(A+\overline{B}+f(0,1))$
$(A+B+f(0,0))$

KARNAUGH MAPS

In a Karnaugh map every possible combination of the binary input variables is represented by a square called a *cell*. The number of squares required is equal to 2^n, where n is the number of variables to cover all possible combinations. There are $2^2 = 4$ squares. A Karnaugh map charts the minterms of ANDed variables.

Taking the simplest case of two variables A and B, the map has four cells, with the four possible combinations in FIG. B-3. Alternatively, the signal values can also be indicated as shown in the right side of FIG. B-3. Drawing larger Karnaugh maps is not really difficult. FIGURE B-4 shows a Kar-

178 Simplifying Digital Logic Circuitry

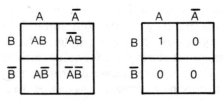

Fig. B-3. Karnaugh maps for two binary variables.

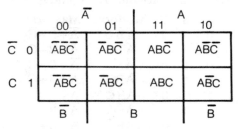

Fig. B-4. A Karnaugh map for three variables.

naugh map for three variables, and FIG. B-5 shows a Karnaugh map for four variables.

The advantage of using a Karnaugh map is that it eliminates any unnecessary inputs from the truth table input patterns used to produce a 1 output, it is quicker to draw and considerably easier to use than a truth table, and it shows you even further simplification of logic circuitry than either the truth table or the Boolean equation can.

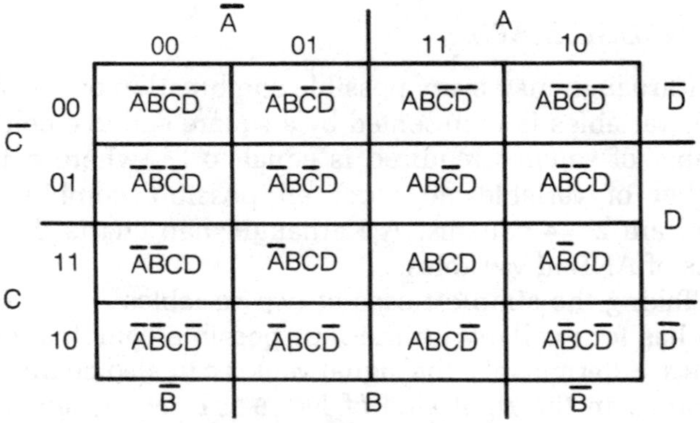

Fig. B-5. A Karnaugh map for four variables.

Simple Operation

For simplicity, assume that two variables, A and B, contain two states, 1 or 0, and a quick check is required on the results of combining A and B as an AND function. A and B are both drawn as separate maps, with respective cell values and annotated by an AND sign. This is shown in FIG. B-6. It is then readily possible to plot the resulting AB map.

In the same way, Karnaugh maps can be used to determine the inverse of a function simply by changing 0's to 1's and 1's to 0's in the individual maps, remembering at the same time this changes AND to OR or vice versa.

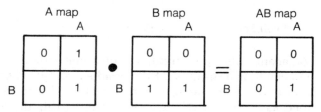

Fig. B-6. AB map derived from A and B maps.

Minimization Techniques

Probably the most useful application of Karnaugh maps is to minimize the number of logic elements necessary to provide a solution to the problem displayed on the map. This involves grouping together adjacent cells on the map with the object of arriving at the simplest statement of the original equation or truth table by graphical rather than mathematical means.

Adjacent cells are cells which differ in only one variable in the AND terms describing the cells. As an example of minimization techniques, suppose the following logic equation involved is:

$$f = \overline{A}B\overline{C} + \overline{A}BC + ABC + AB\overline{C}$$

This, in fact, corresponds to the basic Karnaugh map configuration for three variables without the combinations $\overline{A}\,\overline{B}\,\overline{C}$

and $\overline{A}\overline{B}C$. The resulting Karnaugh map is therefore as shown in FIG. B-7. (The individual cells are designated i, ii, iii, etc., for reference description only; they would not normally be so marked.)

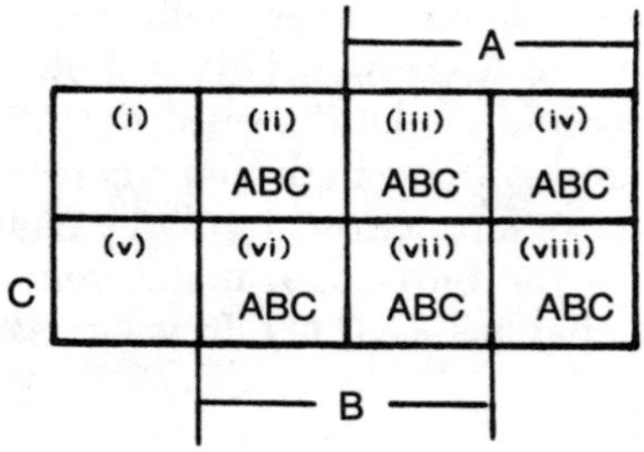

	(i)	(ii) $\overline{A}\overline{B}\overline{C}$	(iii) $A\overline{B}\overline{C}$	(iv) $AB\overline{C}$
C	(v)	(vi) $\overline{A}\overline{B}C$	(vii) $A\overline{B}C$	(viii) ABC

Fig. B-7. 8-cell Karnaugh map for three variables.

Adjacent cells ii and iii differ in only one variable (A and $\overline{A}$), and can thus be grouped. Similar adjacent cells vi and vii differ only in one variable (A and $\overline{A}$ again) and can thus be grouped as shown in FIG. B-8. Having done this the whole map can be defined in simpler terms as follows:

$$f = B\overline{C} + BC$$

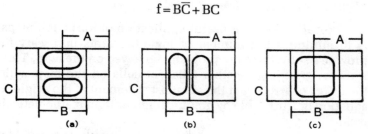

Fig. B-8. Grouping of adjacent cells.

Appendix C

Computer Programming

There are a number of different ways in which you can make a computer understand what it is that you want it to do. You must give it a set of instructions or a program. This can be accomplished by feeding this information into the computer through a keyboard, magnetic tape, or magnetic diskette. However you enter this information, it must be understood by the computer so that it can perform the function it has been asked to complete.

There are several languages that computers understand. Home computers usually base their language on a set of instructions that are written very much like the English language. This type of programming is called BASIC (Beginner's All purpose Symbolic Instruction Code) language. Other languages include *machine language* and *assembly language*. There are, of course, a great many other languages that are used to communicate with computers, but only these will be covered in this section because these are the three most commonly used in understanding computer operation.

BASIC

The program structure of the BASIC language is simply a series of commands stored within the computer's memory and executed only when you type in the command RUN. Storing these commands is accomplished by typing a number in front of the command. Any number of lines may be typed in and they are stored, sequentially, by the computer.

In the BASIC language, a typical program contains an action statement, a control statement that can cause a command to be repeated many times, and a control statement that causes the computer to make a decision to use or skip a part of the instructional program.

Here is a simple program written in BASIC that runs on an IBM PC and is actually a game, asking the player to pick the correct number from 1 to 100:

```
10 READ Q$,D
20 IF D=0 THEN STOP
30 PRINT Q$;
40 INPUT A
50 IF A<D THEN PRINT "TRY HIGHER!":GOTO 20
60 IF A>D THEN PRINT "TRY LOWER!":GOTO 20
70 PRINT "VERY GOOD"
80 GOTO 10
90 DATA "PICK A NUMBER FROM 1 TO 100",67
100 DATA "END",0
```

ASSEMBLY LANGUAGE

In assembly language, symbols are used to represent the commands and are like an abbreviation of the command itself. Where BASIC is three steps removed from the actual language that the machine understands, assembly language is two steps removed.

The following set of programming commands is written in assembly and represents the addition of two numbers in either the direct or extended mode of addressing:

```
LDA A $00F0
ADD A $00F1
STA A $00F2
WAI
```

The first command line tells the computer to load the contents of memory location 00F0 into accumulator A. The second line states "Add the contents of memory location 00F1 to the contents of accumulator A (addition) and place the sum back into accumulator A (replacing the original contents of 00F0 that was originally there)." Next, the computer is told to place the sum that is now in accumulator A into memory location 00F2. Finally, the command WAI, wait for further instructions, is given.

As you can see, this is very close to the actual hexadecimal code that the computer really uses to understand the commands that it is given.

MACHINE LANGUAGE

Machine language is the actual hexadecimal code that the computer understands. The following program is written in machine language and represents the same program of adding two numbers that was just presented using assembly language:

```
B6
00    (LDA A $00F0)
F0

BB
00    (ADD A $00F1)
F1

B7
00    (STA A $00F2)
F2

3E    (WAI)
```

Note: The assembly language shown is for reference only; it would not be written.

Index

A

A/D conversion, 2, 149-150
accumulators, 164
adder/subtractor
  serial, 129-130
  two's complement, 129
adders (see digital adders)
address register, microprocessor, 166
analog memory, 79
analog systems, 1-2
AND gate, 2, 16, 32
  Boolean algebra, 36
  truth table for, 9, 11-12
arithmetic logic unit (ALU), 157-167
  arithmetic operations, 158-161
  cascading, 157-158
  functions of, 157-163
  logic operations, 161-163
  microprocessors and, 163-167
arithmetic operations,
  arithmetic logic unit (ALU), 158-161
assembly language, 183-184

B

BASIC computer programming, 182-183
basic digital concepts, 1-13
  analog systems, 1-2
  binary numbers, 2
  digital systems, 2
  digital terminology, 2
  truth tables, 7-13
binary adders, 123
binary coded decimal (BCD), 6, 85-88
  Diamond code, 91
  Excess Three Code, 89
  Johnson Code, 89
  parity bits, 90-91
  Reflected or Gray Code, 89
  types of code in, 88-89
  weighted codes, 88
binary counters, 134-145
  basic ripple counter, 134-136
  decade counter, 136-139
  divide-by-n counter, 139
  IC binary counters, 144
  IC synchronous counters, 145
  Johnson (twisted ring) counter, 143

Index 185

parallel counters, 140
reversible counter, 136
synchronous counters, 140-143
binary devices, 2
binary numbers, 2-7
  arithmetic, 5
  binary-coded decimal (BCD), 6
  decimal conversion chart, 4
  Gray code conversion, 82
binary subtractors, 127-128
binary-to-decimal conversion, 4
binary/decimal equivalents, 168-172
bipolar transistor, switching characteristics, 25
bistable multivibrators, 102
Boole, George, 30
Boolean Algebra, 10, 30-46, 173-174
  AND gate, 32, 36
  basic logic, 30
  de Morgan's theorem, 45
  enable, 39
  Karnaugh maps, 143, 177-181
  logic symbols, 31
  minimization, 175
  minterms and maxterms, 176
  NAND gate, 32, 37
  NOR gate, 32
  NOT gate, 32
  OR gate, 3G23-35
  problem solving in, 41-44
  theorems in, 45-46
  XOR gate, 37
  YES gate, 32
bootstrap sweep generator, 105
bounce-free switches, 29
buffers, IC, 66

C

cascading arithmetic logic unit (ALU), 157-158
CETOP stardardized symbolism, 14
clocked MOS circuits, 60
clocks (see digital clocks)
CMOS, 58, 60
complementary MOS (CMOS), 58, 60
complex ICs, 67-69
computer programming, 182-184
  assembly language, 183-184
  BASIC language, 182-183
  machine language, 184
control logic, microprocessor, 166
converters
  analog-to-digital (A/D), 149-150
  digital-to-analog, 146-149
cross-coupling, 71
crystal controlled oscillators, 103
current-mode logic (CML), 53

D

D flip-flops, 73-76
D/A conversion, 2, 146-149
data latches (see D flip-flops)
data registers, temporary A&B, 164
decade counter, 136-139
decoders, 110-111
  IC, 114-115
  one-of-sixteen, 115-117
  truth table for, 112
deMorgan's theorem, 45
demultiplexers, 111-114
  one-of-sixteen, 115-117
Diamond code, 91
digital adders, 123-133
  binary adders, 123
  binary subtractors, 127-128
  full-, 124-127
  half- and full-subtractors, 130-133

half-, 123-124
serial adder/subtractor, 129-130
two's complement adder/subtractor, 129
digital clocks, 97-107
  bistable multivibrators, 102
  block diagram of, 107
  crystal controlled oscillators, 103
  frequency division, 107
  IC oscillators, 98-99
  monostable multivibrators, 99-102
  operational amplifier, 97-98
  sweep generators, 103-104
digital families, 69-70
digital logic circuits, 173-181
  Karnaugh maps, 177-181
  minimizing, 175
  minterms and maxterms, 176
  truth tables and Boolean algebra, 173-174
digital systems, 2
digital terminology, 2
diode matrix encoder, 109
diode switches, 26-27
diode-resistor logic network, 48-49
diode-transistor logic (DTL), 50
direct-coupled transistor logic (DCTL), 51-52, 57
display drivers, 120-122
divide-by-n counter, 139
  synchronous, 141
dynamic MOS inverters, 60-61
dynamic MOS NAND gates, 61
dynamic MOS RAM, 82
dynamic MOS shift registers, 155-156

E

electronic switches, 24
emitter-coupled transistor logic (ECTL), 53-54
ENABLE, 39
  truth table, 40
encoders, diode matrix, 109
Excess Three Code, 89
exclusive OR gate (see XOR gate)

F

fan-in/out, 47
FETs
  MOSFETs, 56-58
  switches use of, 26
flip-flops, 71-79
  D, 73-76
  JK master-slave, 77-79
  JK, 76-77
  registers, 84, 151
  RS, 71-73
  symbols for, 19
fractions, various number systems, 95-96
full-adders, 124-127
full-subtractor, 130-133

G

gates (see also logic gates), 2, 28
Gray code, 89
  binary conversion, 82

H

half-adders, 123-124
half-subtractor, 130-133
hexadecimal numbers, 93-95

I

index register, microprocessor, 166
instruction code register, microprocessor, 164
instruction decoder, microprocessor, 166

Index   187

integrated circuits
  binary counters, 144
  buffers, 66
  complex, 67-69
  decoders, 114-115
  minimization of, 63
  multiple-gate, 64-66
  oscillators, 98-99
  RAM, 82-84
  shift registers, 152-155
  standard gates, 63-64
  synchronous counters, 145
inverted parallel logic, 10
inverted series logic, 10

J

JK flip-flops, 76-77
JK master-slave flip-flops, 77-79
Johnson Code, 89
Johnson counter, 143

K

Karnaugh maps, 143, 177-181
  minimization techniques with, 179-181
  operations with, 179

L

LED readouts, 118-119
  display drivers, 120-122
level translators, 64
linear ramp generator, 104
logic, 30
  parallel, 8
  series, 8
logic circuit devices, 47-70
  clocked MOS circuits, 60
  complementary MOS (CMOS), 58, 60
  complex ICs, 67-69
  current-mode logic (CML), 53
  digital families, comparison of, 69-70
  diode-resistor logic networks, 48-49
  diode-transistor logic (DTL), 50
  direct-coupled transistor logic (DCTL), 51-52
  dynamic MOS inverters, 60-61
  dynamic MOS NAND gates, 61
  emitter-coupled transistor logic (ECTL), 53-54
  handling MOS devices, 61-62
  IC buffers, 66
  integrated circuits and minimization, 63
  MOS logic, 58-59
  MOSFETs, 56-58
  multiple-gate ICs, 64-66
  resistor-transistor logic (RTL), 51
  Schmitt trigger, 66-67
  standard IC gates, 63-64
  transistor-transistor logic (TTL), 53, 55, 56
logic gates, 2
  combination of, 11
  cross-coupling, 71
  switches, 15
  YES, 15
logic operations, arithmetic logic unit (ALU), 161-163

M

machine language, 184
master-slave flip-flops, JK, 77-79
mathematical logic (see Boolean algebra)
maxterms, 176
memories, 71, 79-81
  analog, 79
  random-access (RAM), 81-84
  read-only (ROM), 80-81

sample and hold, 79
symbols for, 18
microprocessors, 163-167
  accumulators, 164
  address register, 166
  control logic, 166
  data registers, 164
  index register, 166
  instruction code register, 164
  instruction decoder, 166
  stack pointer, 166
Miller sweep generator, 104
minimization, 175
  Karnaugh maps, 179-181
minterms, 176
monostable multivibrators, 99-102
MOS logic, 58-59
  clocked MOS circuits, 60
  dynamic MOS inverters, 60-61
  dynamic MOS NAND gates, 61
  dynamic MOS shift registers, 155-156
  handling devices using, 61-62
MOSFETs, 56-58
  MOS logic, 58-59
multiple-gate ICs, 64-66
multiplexers, 111-114
multivibrators
  bistable, 102
  monostable, 99-102

N

NAND gate, 2, 16, 32
  Boolean algebra, 37
  dynamic MOS, 61
NOR gate, 2, 17, 32
  truth table for, 10
normal maxterm form, 176
NOT gate (inverter), 2, 16, 32
  truth tables, 7
number systems, 85-96

binary-coded decimal (BCD), 85-88
binary-to-Gray code conversion, 82
binary/decimal equivalents, 168-172
Diamond code, 91
fractions, 95-96
hexadecimal numbers, 93-95
octal numbers, 92-93
parity bits, 90-91

O

octal numbers, 92-93
one-of-sixteen decoder/demultiplexer, 115-117
operational amplifiers, 97-98
OR gate, 2, 17, 32
  Boolean algebra, 33-35
  truth table for, 8-10, 12
oscillators, 98-99
  crystal controlled, 103

P

parallel counters, 140
parallel logic
  inverted, 10
  truth table for, 8
parallel working switches, 23
parity bits, 90-91
positive edge triggered flip-flops (see JK flip-flops)
pulse circuit, Schmitt trigger, 106

R

random-access memories (RAM), 81-84
  dynamic MOS, 82
  typical integrated circuits, 82-84
read-only memories (ROM), 80-81

Reflected Code, 89
registers, 84
　shift (see shift registers)
resistor-transistor logic (RTL), 51
reversible counter, 136
reversible synchronous counters, 141
ripple counter, 134-136
RS flip-flops, 71-73

S

74181 function table, 159
sample and hold memory, 79
Schmitt trigger, 66-67
　pulse circuit, 106
Schottky diodes, 53
　switches use of, 27
serial adder/subtractor, 129-130
series logic
　inverted, 10
　truth table for, 8
series working switches, 23
shift registers, 84, 151
　dynamic MOS, 155-156
　IC, 152-155
silicon controlled rectifier (SCR), 28
Solid State Electronics Theory with Experiments, 26
stack pointer, microprocessor, 166
subtractors (see also digital adders)
　binary, 127-128
　half- and full-, 130-133
sweep generators, 103-104
switches, 18-29
　AND gate, 16
　bipolar transistor, 25
　bounce-free, 29
　digital logic gates, 15
　diode, 26-27
　electronic, simple, 24
　FETs as, 26
　functions of, 18
　improving transistor switch-off times, 26
　NAND gate, 16
　NOR gate, 17
　NOT gate (inverter), 16
　OR gate, 17
　Schottky diodes, 27
　series and parallel working, 23
　solving equations with, 20-22
　thyristors, 28
　truth tables and, 13, 19
　unijunction transistors, 27
　XOR gate, 17
symbols, 14-29
　flip-flops, 19
　logic, Boolean equations, 31
　memory, 18
synchronous counters, 140-143
　divide-by-n counter, 141
　integrated circuits, 145
　reversible, 141

T

thyristors, 28
transistor-transistor logic (TTL), 53, 55, 56
transistors
　bipolar, switching characteristics, 25
　improving switch-off times, 26
　Schottky, 27
　unijunction, 27
triacs, 28
triggers, Schmitt, 66-67
truth tables, 7-13, 173-174
　AND gate, 9, 11-12
　decoder, 112
　enable, 40
　full-adder, 125
　logic gate combinations, 11
　NOR gate, 10

OR gate, 8-10, 12
parallel logic, 8
plotting, 12
series logic, 8
switches and, 13, 19
XOR gate, 39
twisted ring counter, 143
two's complement adder/subtractor, 129

U

unijunction transistors, 27
US MIL standardized symbolism, 14

W

weighted code, BCD, 88

X

XOR gate, 17
 Boolean algebra, 37
 truth table, 39

Y

YES gate, 15, 32